Asheville-Buncombe
Technical Community College
Learning Resources Center
340 Victoria Rd.
Asheville, NC   28801

Wireless Networking Handbook

Jim Geier

Asheville-Buncombe
Technical Community College
Learning Resources Center
340 Victoria Rd.
Asheville, NC 28801

New Riders Publishing, Indianapolis, Indiana

Wireless Networking Handbook

By Jim Geier

Published by:
New Riders Publishing
201 West 103rd Street
Indianapolis, IN 46290 USA

Copyright © 1996 by New Riders Publishing

Printed in the United States of America 1 2 3 4 5 6 7 8 9 0

Library of Congress Cataloging-in-Publication Data

CIP data available upon request

Warning and Disclaimer

This book is designed to provide information about wireless networking. Every effort has been made to make this book as complete and as accurate as possible, but no warranty or fitness is implied.

The information is provided on an "as is" basis. The author and New Riders Publishing shall have neither liability nor responsibility to any person or entity with respect to any loss or damages arising from the information contained in this book or from the use of the disks or programs that may accompany it.

Publisher	Don Fowley
Publishing Manager	Emmett Dulaney
Marketing Manager	Mary Foote
Managing Editor	Carla Hall

New Riders Publishing

The staff of New Riders Publishing is committed to bringing you the very best in computer reference material. Each New Riders book is the result of months of work by authors and staff who research and refine the information contained within its covers.

As part of this commitment to you, the NRP reader, New Riders invites your input. Please let us know if you enjoy this book, if you have trouble with the information and examples presented, or if you have a suggestion for the next edition.

Please note, though: New Riders staff cannot serve as a technical resource for wireless networking or for questions about software- or hardware-related problems.

If you have a question or comment about any New Riders book, there are several ways to contact New Riders Publishing. We will respond to as many readers as we can. Your name, address, or phone number will never become part of a mailing list or be used for any purpose other than to help us continue to bring you the best books possible. You can write us at the following address:

New Riders Publishing
Attn: Publisher
201 W. 103rd Street
Indianapolis, IN 46290

If you prefer, you can fax New Riders Publishing at (317) 581-4670.

You can also send electronic mail to New Riders at the following Internet address:

jbelbot@newriders.mcp.com

NRP is an imprint of Macmillan Computer Publishing. To obtain a catalog or information, or to purchase any Macmillan Computer Publishing book, call (800) 428-5331.

Thank you for selecting *Wireless Networking Handbook*!

Trademark Acknowledgments

Dedication

I'd like to express gratitude to my wife, Debbie, for reviewing and editing drafts of the book. Her assistance made the process much easier to bear. Also, many thanks to the companies who provided me information about their products and allowed the reprinting of their case studies and product descriptions.

Finally, I'd like to dedicate this book to my sons, Brian and Eric Geier, as well as Evan and Jared Ream.

Contents at a Glance

Table of Contents

Part II: Analyzing the Need for Wireless Networks

Part III: Implementing and Supporting Wireless Networks

Part IV: Appendixes

PART I

Wireless Network Technologies and Standards

1. Introduction to Wireless Networking

2. Wireless Local Area Networks (LANs)

3. Wireless Metropolitan Area Networks (MANs)

4. Wireless Wide Area Networks (WANs)

Introduction to Wireless Networking

Many organizations utilize traditional wire-based networking technologies to establish connections among computers. These technologies fall into the following three categories:

○ Local area networks (LANs)

○ Metropolitan area networks (MANs)

○ Wide area networks (WANs)

LANs support the sharing of applications and printers, transfer of files, and sending e-mail within a room or building. Today, the industry standard for LANs is ethernet technology with 10baseT Category 5 twisted-pair wiring. MANs, which can cover the size of a college campus or large city, interconnect LANs by using protocols such as FDDI (Fiber Distributed Data Interface) and depend on leased circuits and optical fiber for transmission of the data. WANs, on the other hand, utilize telephone circuits, leased lines, and private circuits to support worldwide networking by using circuit and packet switching protocols.

Traditional networking technologies offer tremendous capabilities from an office, hotel room, or home. Activities such as communicating via e-mail with someone located in a faraway town or conveniently accessing product information from the World Wide Web are the result of widespread networking. But, limitations to networking through the use of wire-based systems exist because you cannot utilize these network services unless you are physically connected to a LAN or a telephone connection.

Over the last thirty years, researchers and companies have been busy developing protocols and systems that provide wireless connectivity for LANs, MANs, and WANs. This work has not been easy and has met much resistance from end users. Today, though, products are available that fit into all categories of networks and satisfy the need for mobility. This chapter introduces wireless networking by describing the following:

- ○ The history of wireless networks
- ○ Wireless network architecture
- ○ The benefits of wireless networking
- ○ Concerns surrounding the implementation and use of wireless networks
- ○ The wireless network market
- ○ The future of wireless networks

The History of Wireless Networks

The first indication of wireless networking dates back to the 1800s and earlier. Indians, for example, sent information to each other via smoke signals from a burning fire. The sender would simply wave a deer skin over the fire—in sequences similar to Morse code—to send messages warning others that a war was imminent, or just to say, "I'll be late for dinner." This smoke signal system was a true network. People manning intermediate fires would relay messages if a great distance separated the source of the message and the destination. The world has seen much progress since those days. Messengers on horseback eventually became a more effective means of transferring information, while later the telephone made communications much easier and faster.

Prior to the nineteenth century, scientists thought light was the only wavelength component of the electromagnetic spectrum. During the nineteenth and twentieth centuries, researchers learned that the spectrum actually consists of longer wavelengths (lower frequencies) as well. Experiments showed that lower frequencies, such as radio waves and infrared light, could be sent through the air with moderate amounts of transmit power and easy-to-manufacture antennas. As a result, companies began building radio transmitters and receivers, making public and private radio communications, television, and wireless networking possible.

Network technologies and radio communications were brought together for the first time in 1971 at the University of Hawaii as a research project called ALOHANET. The ALOHANET system enabled computer sites at seven campuses spread out over four islands to communicate with the central computer on Oahu without using existing, unreliable, expensive phone lines. ALOHANET offered bi-directional communications, in a star topology, between the central computer and each of the remote stations. The remote stations had to communicate with each other via the centralized computer. ALOHANET became popular among network researchers because of the unique combination of packet switching and broadcast radio. The U.S. military embraced the technology, and DARPA (Defense Advanced Research Projects Agency) began testing wireless networking to support tactical communications in the battlefield. Because of limited spectrum allocations, radio-based networks could only deliver very low data rates. Research done at the University of Hawaii and DARPA, however, helped pave the way for the development of the initial ethernet technology, as well as fueled the development of radio-based networks available today.

The advent of the wired ethernet technology steered many commercial companies away from radio-based networking components toward the production of ethernet-related products. Companies did not mind running cable throughout and between their facilities to take advantage of ethernet's whopping 10 Mbps data rates. In the eighties, amateur radio hobbyists, "hams," kept radio networking alive within the U.S. and Canada by designing and building Terminal Node Controllers (TNCs) to interface their computers through ham radio equipment (see fig. 1.1). These TNCs act much like a telephone modem, converting the computer's digital signal into one that a ham radio can modulate and send over the airwaves by using a packet switching technique. In fact, the American Radio Relay League (ARRL) and the Canadian Radio Relay League (CRRL) have been sponsoring the Computer Networking Conference since the early eighties in order to provide a forum for the development of wireless WANs. Thus, hams have been utilizing wireless networking for years, much longer than the commercial market.

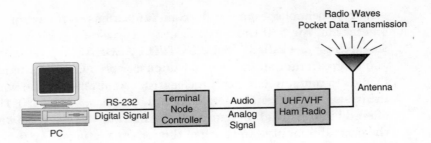

Figure 1.1
The ham radio terminal node controllers.

In 1985, the Federal Communications Commission (FCC) made the commercial development of radio-based LAN components possible by authorizing the public use of the Industrial, Scientific, and Medical (ISM) bands. This band of frequencies resides between 902 MHz and 5.85 GHz, just above the cellular phone operating frequencies. The ISM band is very attractive to wireless network vendors because it provides a part of the spectrum upon which to base their products, and end users do not have to obtain FCC licenses to operate the products. The ISM band allocation has had a dramatic effect on the wireless industry, prompting the development of wireless LAN components.

During the late eighties, the decreasing size of computers from desktop machines to laptops allowed employees to take their computers with them around the office and on business trips. Computer companies then scrambled to develop products that would support wireless connectivity methods. In 1990, NCR began shipping WaveLAN, one of the first wireless LAN adapters for PCs. Motorola was also one of the initial wireless LAN vendors with a product called Altair. These early wireless network adapters had limited network drivers, but soon worked with almost any network operating system. Rapidly, companies such as Proxim, Xircom, Windata, and others began shipping their products as well. These initial companies were pioneers in the wireless networking arena. They felt their wireless products would feed a market desperately wanting to meet mobility needs. Network managers and system administrators, however, did not trust the technology enough to purchase the pricy wireless adapters. The WaveLAN network adapters, for example, initially listed for $1,400 each. Vendors quickly found that they had to reduce prices. Businesses were afraid these new wireless products lacked security, were too slow (at least slower than ethernet), and were not standardized. Over time, though, vendors began incorporating encryption to protect data transmissions, and operation at higher frequencies increased the bandwidth to near-ethernet speeds. The market also saw a substantial drop in price to $300–$500 per card. The lack of standards, however, limited widespread use of wireless LAN products.

The current depressed state of the wireless LAN market should change as standards mature. The Institute for Electrical and Electronic Engineers (IEEE) 802 Working Group, responsible for the development of LAN standards such as ethernet and token ring, initiated the 802.11 Working Group to develop a standard for wireless LANs. This group began operations in the late eighties under the chairmanship of Vic Hayes, an engineer from NCR. At the present time, 802.11 is still working on the standard. A final standard should pass in 1997.

End users and network managers have had a difficult time showing a positive business case for purchasing wireless LAN components in the office unless there is a requirement for mobility. Most sales of wireless LAN adapters to date have been in healthcare and financial environments. Sensing a bleak market for wireless LAN products, wireless LAN vendors began equipping their wireless LAN components in 1995 with directional antennas to facilitate point-to-point connections between buildings located within the same metropolitan area. These wireless MAN products satisfy a widespread need—the capability to connect facilities where traditional cable installation and leased circuits are costly. The sales of these wireless MAN products have been favorable.

The most widely accepted wireless network connection, though, has been wireless WAN services, which began surfacing in the early nineties. Companies such as ARDIS and RAM Mobile Data were first in selling wireless connections between portable computers, corporate networks, and the Internet. This service enables employees to access e-mail and other information services from their personal appliance without using the telephone system when meeting with customers, traveling in the car, or staying in a hotel room.

Narrowband Personal Communications Services (PCS), a spectrum allocation located at 1.9 GHz, is a new wireless communications technology offering wireless access to the World Wide Web, e-mail, voice mail, and cellular phone service. Vice President Al Gore kicked off the FCC PCS auction in 1995 by selling 30-MHz licenses to television and telephone companies. The total take for 1995 was $7.7 billion. The U.S. government expects to raise $15 billion from the auctioning during 1996.

Because of PCS, the wireless industry is quickly gaining momentum. As a result, a vast number of wireless networking products should appear on the market in 1997. SkyTel began shipping the first PCS product in 1996, a pocket-sized two-way pager, which can receive pages as well as respond.

Wireless Network Architecture

In general, networks perform many functions to transfer information from source to destination.

1. The medium provides a bit pipe (path for data to flow) for the transmission of data.

2. Medium access techniques facilitate the sharing of a common medium.

3. Synchronization and error control mechanisms ensure that each link transfers the data intact.

4. Routing mechanisms move the data from the originating source to the intended destination.

A good way to depict these functions is to specify the network's architecture. This architecture describes the protocols, major hardware, and software elements that constitute the network. A network architecture, whether wireless or wired, may be viewed in two ways, logically and physically.

Logical Architecture of a Wireless Network

A *logical architecture* defines the network's protocols—rules by which two entities communicate. People observe protocols every day. Individuals participating in a business meeting, for example, interchange their ideas and concerns while they avoid talking at the same time. They also rephrase a message if no one understands it. Doing so ensures a well-managed and effective means of communication. Likewise, PCs, servers, routers, and other active devices must conform to very strict rules to facilitate the proper coordination and transfer of information.

One popular standard logical architecture is the 7-layer Open System Interconnection (OSI) Reference Model, developed by the International Standards Organization (ISO). OSI specifies a complete set of network functions, grouped into layers. Figure 1.2 illustrates the OSI Reference Model.

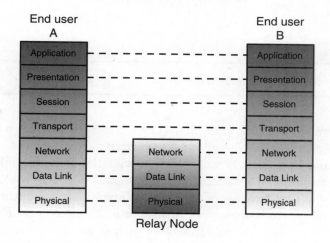

Figure 1.2
The Open System Interconnection Reference Model.

The OSI layers provide the following network functionality:

○ **Layer 7—Application layer.** Establishes communications with other users and provides services such as file transfer and e-mail to the end users of the network.

○ **Layer 6—Presentation layer.** Negotiates data transfer syntax for the application layer and performs translations between different data types, if necessary.

○ **Layer 5—Session layer.** Establishes, manages, and terminates sessions between applications.

○ **Layer 4—Transport layer.** Provides mechanisms for the establishment, maintenance, and orderly termination of virtual circuits, while shielding the higher layers from the network implementation details.

○ **Layer 3—Network layer.** Provides the routing of packets from source to destination.

○ **Layer 2—Data Link layer.** Ensures synchronization and error control between two entities.

○ **Layer 1—Physical layer.** Provides the transmission of bits through a communication channel by defining electrical, mechanical, and procedural specifications.

NOTE

Each layer of OSI supports the layers above it.

Does a wireless network offer all OSI functions? No. As shown in figure 1.3, wireless LANs and MANs function only within the Physical and Data Link layers, which provide the medium, link synchronization, and error control mechanisms. Wireless WANs provide these first two layers, as well as Network Layer routing. In addition to the wireless network functions, a complete network architecture needs to include functions such as end-to-end connection establishment and application services.

Physical Architecture of a Wireless Network

The physical components of a wireless network implement the Physical, Data Link, and Network Layer functions (see fig. 1.4). The network operating system (NOS) of a network, such as Novell Netware, supports the shared use of applications, printers, and disk space. The NOS, located on client and server machines, communicates with the wireless Network Interface Card (NIC) via driver software, enabling applications to utilize the wireless network for data transport. The NIC prepares data signals for propagation from the antenna through the air to the destination comprised of the same set of components.

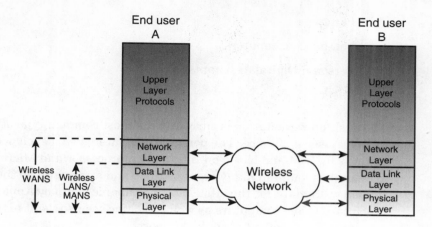

Figure 1.3

The wireless network logical architecture.

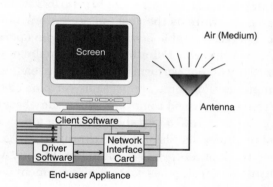

Figure 1.4

The physical components of a wireless network.

End-User Appliances

As with any system, there must be a way for users to interface with applications and services. Whether the network is wireless or wired, an *end-user appliance* is a visual interface between the user and the network. Following are the classes of user appliances:

○ Desktop workstations

○ Laptops

- ○ Palmtops
- ○ Pen-based computers
- ○ Personal Digital Assistants (PDA)
- ○ Pagers

The desktop workstation is currently the most common type of network user appliance. The personal computer (PC), developed initially by IBM and based on Intel's microprocessor and Microsoft's Windows, is found in many organizations and appears to be the industry standard for the office. Some companies employ Apple's Macintosh (Mac) computers as well. The Mac seems to suit artisans because of its excellent graphics support.

Smaller computers have been effective in satisfying portable computing needs of business executives and other professionals. Laptops, which measure roughly 8×10×3 inches, can run the same type of software as the desktop computers, but fit in a briefcase and include rechargeable batteries to sustain operations where electricity is not present. Palmtops fit in the palm of your hand, but they generally do not perform as well as the leading laptop and desktop computers. For some applications, such as electronic patient record keeping, pen-based computers are handy because they enable you to enter data into a portable device via a pen. More and more mobile professionals also are turning to PDAs that enable them to keep track of contacts, schedules, and tasks. Apple's Newton PDA, for instance, combines contact management software and electronic book creation software for salespeople. Pagers are also available as user appliances. Pagers get your attention when someone calls a special telephone number. These devices, however, are more than just beepers. Some pagers are now capable of receiving and sending limited alphanumeric text due to narrowband PCS.

Because wireless network appliances are often put into the hands of mobile people who work outside, the appliance must be tough enough to resist damage resulting from dropping, bumping, moisture, and heat. Some companies now offer more durable versions of the portable computer. Itronix, for example, sells the X-C 6000 Cross Country 486 portable computer. The X-C 6000's case is built from strong, lightweight magnesium and includes an elastomer covering that protects the unit from weather and shock. The unit is impervious to rain, beverage spills, and other work environment hazards.

Some of these vendors also produce pen-based and handheld computers well-suited for wireless applications. Telxon has a PTC-1184 full-screen 486 pen-based computer that combines pen-based technology with bar code scanning and AIRONET's MicroCellular radio-based wireless network interface card. This system makes it possible to use the unit in environments such as hospitals, factory floors, and warehouses.

Network Software

A wireless network supports the NOS and its applications, such as word processing, databases, and e-mail, enabling the flow of data between all components. NOSs provide file and print services, acting as a platform for user applications. Many NOSs are server-oriented, as shown in figure 1.5, where the core software resides on a high-performance PC. A client, located on the end user's appliance, includes server software that directs the user's command to the local computer resources, or puts it out onto the network to another computer. Some wireless networks may also contain middleware that interfaces mobile applications to the wireless network hardware.

Wireless Network Interface

Computers process information in digital form, with low direct current (DC) voltages representing data ones and zeros. These signals are optimum for transmission within the computer, not for transporting data through wired or wireless media. A wireless network interface couples the digital signal from the end-user appliance to the wireless medium, which is air, to enable an efficient transfer of data between sender and receiver. This process includes the modulation and amplification of the digital signal to a form acceptable for propagation to the receiving location. *Modulation* is the process of translating the baseband digital signal to a suitable analog form. This process is very similar to the common telephone modem, which converts a computer's digital data into an analog form within the 4 KHz limitation of the telephone circuit. The wireless modulator translates the digital signal to a frequency that propagates well through the atmosphere. Wireless networks employ modulation by using radio waves and infrared light. *Amplification* raises the amplitude of the signal so it will propagate a greater distance. Without amplification, you would have a difficult time talking to a crowd of 1,500 people outside in an open area. But, add amplification, such as a PA system, and everyone can hear you.

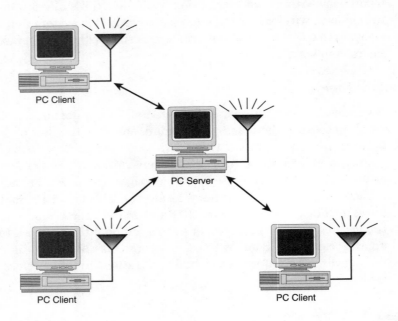

Figure 1.5
The server-based network operating system.

N O T E

Chapter 2, "Wireless Local Area Networks (LANs)," covers modulation techniques in more detail.

The wireless network interface also manages the use of the air through the operation of a communications protocol. For synchronization, wireless networks employ a carrier sense protocol similar to the common ethernet standard. This protocol enables a group of wireless computers to share the same frequency and space. As an analogy, consider a room of people engaged in a single conversation in which each person can hear if someone speaks. This represents a fully connected bus topology (where everyone communicates using the same frequency and space) that ethernet and wireless networks, especially wireless LANs, utilize. To avoid having two people speak at the same time, you should wait until the other person has finished talking. Also, no one should speak unless the room is silent. This simple protocol ensures only one person speaks at a time, offering a shared use of the communications medium. Wireless networks use carrier sense protocols and operate in a similar fashion, except the communications are by way of radio signals or infrared light. Figure 1.6 illustrates the generic carrier sense protocol.

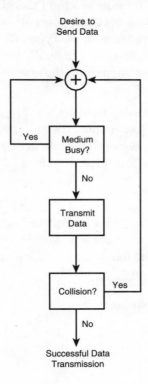

Figure 1.6
The operation of the carrier sense protocol.

Wireless networks handle error control by having each station check incoming data for altered bits. If the destination station does not detect errors, it sends an acknowledgment back to the source station. If the station detects errors, the data link protocol ensures that the source station resends the packet. To continue the analogy, consider two people talking to each other outside. If one person is speaking and a disruption occurs, such as a plane flying overhead, the dialog might become distorted. As a result, the listener asks the speaker to repeat a phrase or two. The wireless network interface generally takes the shape of a wireless NIC or an external modem that facilitates the modulator and communications protocols. These components interface with the user appliance via a computer bus, such as ISA (Industry Standard Architecture) or PCMCIA (Personal Computer Memory Card International Association). The ISA bus comes standard in most desktop PCs. Many portable computers have PCMCIA slots that accept credit card-sized NICs. PCMCIA specifies three interface sizes, Type I (3.3 millimeters), Type II (5.0 millimeters), and Type III (10.5 millimeters). Some companies also produce wireless components that connect to the computer via the RS-232 serial port.

The interface between the user's appliance and NIC also includes a software driver that couples the client's application or NOS software to the card. Several de facto driver standards exist, including ODI (Open Data-Link Interface) and NDIS.

Antenna

The antenna radiates the modulated signal through the air so that the destination can receive it. Antennas come in many shapes and sizes and have the following specific electrical characteristics:

- ○ Propagation pattern
- ○ Radiation power
- ○ Gain
- ○ Bandwidth

The *propagation pattern* of an antenna defines its coverage. A truly omnidirectional antenna transmits its power in all directions, whereas a directional antenna concentrates most of its power in one direction. Figure 1.7 illustrates the differences. Radiation power is the output of the radio transmitter. Most wireless network devices operate at less than 5 watts of power.

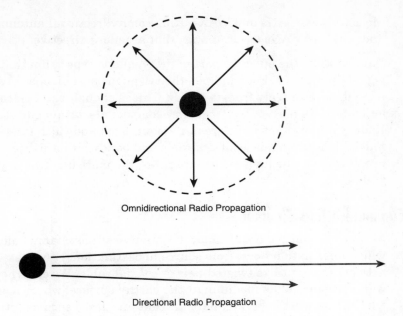

Omnidirectional Radio Propagation

Directional Radio Propagation

Figure 1.7

The omnidirectional versus directional antennas.

A directional antenna has more gain (degree of amplification) than the omnidirectional type and is capable of propagating the modulated signal farther because it focuses the power in a single direction. The amount of gain depends on the directivity of the antenna. An omnidirectional antenna has a gain equal to one; that is, it doesn't focus the power in any particular direction. A directional antenna, however, is considered to add gain (amplification) to the signal in certain directions. Consider the example of watering your lawn with a garden hose. Attach a circular sprayer to the end of the hose and turn the water on. The water pressure is divided among many directions, and the resulting water spray will reach seven or eight feet. If a more directive nozzle is attached to the hose, the spray might reach twenty feet because the device concentrates the water pressure in one direction. Similarly, the combination of transmit power and gain of an antenna defines the distance the signal will propagate. Long-distance transmissions require higher power and directive radiation patterns. With wireless networks, these signals are relatively low power, typically one watt or less.

Most wireless LANs and WANs utilize omnidirectional antennas, and wireless MANs use antennas that are more directive.

Bandwidth is the effective part of the frequency spectrum that the signal propagates. For example, the telephone system operates over a bandwidth roughly from 0–4 KHz. This is enough bandwidth to accommodate most of the frequency components within our voices. Radio wave systems have greater amounts of bandwidth located at much higher frequencies. Data rates and bandwidth are directly proportional—the higher the data rates, the more bandwidth you will need.

The Communications Channel

All information systems employ a communications channel along which information flows from source to destination. Ethernet networks may utilize twisted-pair or coaxial cable. Wireless networks use air as the medium. At the earth's surface, where most wireless networks operate, pure air contains gases such as nitrogen and oxygen. This atmosphere provides an effective medium for the propagation of radio waves and infrared light. Rain, fog, and snow, however, can increase the amount of water molecules in the air and can cause significant attenuation to the propagation of modulated wireless signals. Smog clutters the air, adding attenuation to the communications channel as well. *Attenuation* is the decrease in the amplitude of the signal, and it limits the operating range of the system. The ways to combat attenuation are to either increase the transmit power of the wireless devices, which in most cases is limited by the FCC, or to incorporate special amplifiers called repeaters that receive attenuated signals, revamp them, and transmit downline to the end station or next repeater.

Benefits of Wireless Networks

Companies can realize the following benefits by implementing wireless networks:

○ Mobility

○ Ease of installation in difficult-to-wire areas

○ Reduced installation time

○ Increased reliability

○ Long-term cost savings

Mobility

User mobility indicates constant physical movement of the person and their network appliance. Many jobs require workers to be mobile, such as inventory clerks, healthcare workers, policemen, emergency care specialists, and so on. Wireline networks require a physical tether between the user's workstation and the network's resources, which makes access to these resources impossible while roaming about the building or elsewhere. As an analogy, consider talking on a wired phone having a cord connecting the handset to the telephone base station. You can utilize the phone only within the length of its cord. With a wireless cellular phone, however, you can walk freely within your office, home, or even talk to someone while driving a car. Wireless networking offers mobility to its users much like the wireless phone, providing a constant connection to information on the network. This connection can be extremely useful if you are at a customer's site discussing a new product, delivering emergency care to a crash victim, or in a hotel room sending and receiving e-mail. You cannot become mobile unless you eliminate the wire through the use of wireless networking.

Installation in Difficult-to-Wire Areas

The implementation of wireless networks offers many tangible cost savings when performing installations in difficult-to-wire areas. If rivers, freeways, or other obstacles separate buildings you want to connect (see fig. 1.8), a wireless MAN solution may be much more economical than installing physical cable or leasing communications circuits such as T1 service or 56 Kbps lines. Some organizations spend hundreds, thousands, or even millions of dollars to install physical links with nearby facilities. If you are facing this type of installation, consider wireless networking as an alternative. The deployment of wireless networking in these situations costs thousands of dollars, but will result in a definite cost savings in the long run.

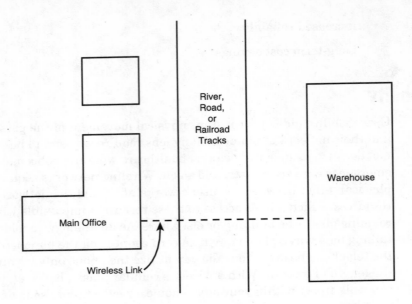

Figure 1.8
A difficult-to-wire situation.

The asbestos found in older facilities is another problem that many organizations encounter. The inhalation of asbestos particles is extremely hazardous to your health; therefore, you must take great care when installing network cabling within these areas. When taking necessary precautions, the resulting cost of cable installations in these facilities can be prohibitive. Some organizations, for example, remove the asbestos, making it safe to install cabling. This process is very expensive because you must protect the building's occupants from breathing the asbestos particles agitated during removal. The cost of removing asbestos covering just a few flights of stairs can be tens of thousands of dollars. Obviously, the advantage of wireless networking in asbestos-contaminated buildings is that you can avoid the asbestos removal process, resulting in tremendous cost savings.

In some cases, it might be impossible to install cabling. Some municipalities, for example, may restrict you from permanently modifying older facilities with historical value. This could limit the drilling of holes in walls during the installation of LAN cabling and network outlets. In this situation, a wireless LAN might be the only solution. Right-of-way restrictions within cities and counties may also block the digging of trenches in the ground to lay optical fiber for the interconnection of networked sites. Here, a wireless MAN or WAN might be the only alternative.

Reduced Installation Time

The installation of cabling is often a time-consuming activity. For LANs, installers must pull twisted-pair wires above the ceiling and drop cables through walls to network outlets that they must affix to the wall. These tasks can take days or weeks, depending on the size of the installation. The installation of optical fiber between buildings within the same geographical area consists of digging trenches to lay the fiber or pulling the fiber through an existing conduit. You might need weeks or possibly months to receive right-of-way approvals and dig through ground and asphalt. The deployment of wireless LANs, MANs, or WANs greatly reduces the need for cable installation, making the network available for use much sooner. Thus, many countries lacking a network infrastructure have turned to wireless networking as a method of providing connectivity among computers without the expense and time associated with installing physical media.

Increased Reliability

A problem inherent to wired networks is the downtime due to cable faults. Moisture erodes metallic conductors. These imperfect cable splices can cause signal reflections that result in unexplainable errors. The accidental cutting of cables can also bring a network down quickly. Water intrusion can also damage communications lines during storms. These problems interfere with the users' ability to utilize network resources, causing havoc for network managers. The advantage of wireless networking, then, is experiencing fewer problems because less cable is used.

Long-Term Cost Savings

Companies reorganize, resulting in the movement of people, new floor plans, office partitions, and other renovations. These changes often require re-cabling the network, incurring both labor and material costs. In some cases, the re-cabling costs of organizational changes are substantial, especially with large enterprise networks. A reorganization rate of 15% each year can result in yearly reconfiguration expenses as high as $250,000 for networks having 6,000 interconnected devices. The advantage of wireless networking is again based on the lack of cable—you can move the network connection by simply relocating an employee's PC.

Wireless Network Concerns

The benefits of a wireless network are certainly welcomed by companies and organizations. Network managers and engineers should be aware, however, of the following concerns that surround the implementation and use of wireless networking:

- ○ Radio signal interference
- ○ Power management
- ○ System interoperability
- ○ Network security
- ○ Installation issues
- ○ Health risks

Radio Signal Interference

The purpose of radio-based networks is to transmit and receive signals efficiently over airwaves. This process, though, makes these systems vulnerable to atmospheric noise and transmissions from other systems. In addition, these wireless networks could interfere with other radio wave equipment. As shown in figure 1.9, interference may be inward or outward.

Inward Interference

Most of us have experienced radio signal interference while talking on a wireless telephone, watching television, or listening to a radio. Someone close by might be communicating with another person via a short-wave radio system, causing harmonic frequencies that you can hear while listening to your favorite radio station. Or, a remote control car can cause static on a wireless phone while you are attempting to have a conversation. These types of interference might also disturb radio-based wireless networks in the form of inward interference.

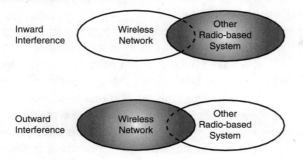

Figure 1.9
Inward and outward interference.

A radio-based LAN, for example, can experience some inward interference either from the harmonics of transmitting systems or other products using ISM-band frequencies in the local area. Microwave ovens operate in the S band (2.4 GHz) that many wireless LANs transmit and receive. These signals result in delays to the user by either blocking transmissions from stations on the LAN or causing bit errors to occur in data being sent. These types of interference can limit the areas in which you can deploy a wireless network. As an illustration, when deploying a wireless network at a site located in Washington, D.C., along the Potomac River, a company occasionally experienced a great deal of delay from stations located on the side of the building facing the river. The implementation team found, through radio propagation tests, that a military base on the opposite side of the river was periodically transmitting a radio signal. The interfering signal was strong enough for the LAN stations to misinterpret it as data traffic, forcing the stations to wait an inefficient period of time.

N O T E

To make matters worse, most radio-based products operate within the public, license free, ISM bands. These products do not require users to obtain FCC licenses, which

continues

means the FCC does not manage the use of the products. If you experience interference within the ISM band resulting from another product operating within that band, you have no recourse. The FCC is not committed to step in and resolve the matter, leaving you with the choice of dealing with delays the interference causes or looking for a different technology to support your needs.

Interference with radio-based networks is not as bad as it might seem. The products using the ISM bands incorporate spread spectrum modulation that limits the amount of damage an interfering signal causes. The spread spectrum signal covers a wide amount of bandwidth, and a typical narrow bandwidth interference only affects a small part of the information signal, resulting in few or no errors. Thus, spread spectrum-type products are highly resistant to interference. Narrowband interference with signal-to-interference ratios of less than 10 dB does not usually affect a spread spectrum transmission. Wideband interference, however, can have damaging effects on any type of radio transmission. The primary source of wideband interference is domestic microwave ovens that operate in the 2.4 GHz band. The typical microwave oven operates at 50 pulses per second and sweeps through frequencies between 2400 and 2450 MHz, corrupting the wireless data signal if within 50 feet of the interfering source. Other interference may result from elevator motors, duplicating machines, theft protection equipment, and cordless phones.

Outward Interference

Inward interference is only half of the problem. The other half of the issue, outward interference, occurs when a wireless network's signal disrupts other systems, such as adjacent wireless LANs, navigation equipment on aircraft, and so on. This disruption results in the loss of some or all of the system's functionality. Interference is uncommon with ISM band products because they operate on such little power. The transmitting components must be very close and operating in the same bandwidth for either one to experience inward or outward interference.

Techniques for Reducing Interference

When dealing with interference, you will want to coordinate the operation of radio-based wireless network products with your company's frequency management organization, if one exists. This will avoid potential interference problems. In fact, the coordination with frequency management officials is mandatory before operating radio-based wireless devices of any kind on a U.S. military base. The military does not follow the same frequency allocations issued by the FCC. The FCC deals with commercial sectors of the U.S., and the military has their own frequency management process. You must obtain special approvals from the government to operate ISM-based products on military bases because they may interfere with some of their systems. The approval process can take several months to complete.

Another tip, especially if no frequency management organization exists within your company, is to run some tests to determine the propagation patterns within your building. These tests let you know if existing systems may interfere with, and thus block and cause delay to, your network. You will also discover whether your signal will disturb other systems. See Chapter 8, "Designing a Wireless Network," for details on ways to perform propagation tests (site survey).

Power Management

If you are using a portable computer in an automobile, performing an inventory in a warehouse, or caring for patients in a hospital, it might be cumbersome or impossible to plug your computer into an electrical outlet. Thus, you will be dependent on the computer's battery. The extra load of the wireless NIC in this situation can significantly decrease the amount of time you have available to operate the computer before needing to recharge the batteries. Your operating time, therefore, might decrease to less than an hour if you access the network often.

To counter this problem, vendors implement power management techniques in their PCMCIA format wireless NICs. Proxim's wireless LAN product, RangeLAN2/PCMCIA, for example, maximizes power conservation. RangeLAN2 accommodates advanced power management features found in most portable computers. Without power management, radio-based wireless components normally remain in a receptive state waiting for any information. Proxim incorporates two modes to help conserve power: the Doze Mode and the Sleep Mode. The Doze Mode, which is the default state of the product, keeps the radio off most of the time and wakes up periodically to determine if any messages await in a special mailbox. This mode alone utilizes approximately 50 percent less battery power. The Sleep Mode causes the radio to remain in a transmit-only standby mode. In other words, the radio wakes up and sends information if necessary, but is not capable of receiving any information. Other products offer similar power management features.

System Interoperability

When implementing an ethernet network, network managers and engineers can deploy NICs from a variety of vendors on the same network. Because of the stable IEEE 802.3 standard that specifies the protocols and electrical characteristics that manufacturers must follow for ethernet, these products all speak exactly the same language. This uniformity allows you to select products meeting your requirements at the lowest cost from a variety of manufacturers. Today, this is not possible with most wireless network products, especially wireless LANs and MANs. The selection of these wireless products is predominantly single vendor, sole-source acquisitions. Products from one vendor will not interoperate with those from a different company. This raises a problem when deploying the network. Once you decide to buy a particular brand of wireless network component, you must continue to purchase that brand to ensure that the components can talk the same language as the existing ones. Putting yourself in this situation is risky. What happens if your wireless vendor decides to discontinue the product you chose?

As mentioned earlier, the solution to this problem, at least for wireless LANs, is very near. The IEEE 802.11 Working Group plans to issue final standards for wireless LANs by 1997. Wireless LAN vendors should embrace the standard because they are active in the standard development process. Shifting their products to the standard will be easy for them.

Network Security

Network security refers to the protection of information and resources from loss, corruption, and improper use. Are wireless networks secure? Among businesses considering the implementation of a wireless system, this is a common and very important question. To answer this question, you must consider the functionality a wireless network performs. As described earlier, a wireless network provides a bit pipe, consisting of a medium, synchronization, and error control that supports the flow of data bits from one point to another. This setup corresponds to the lowest levels of the network architecture and does not include other functions such as end-to-end connection establishment or login services. Therefore, the only security issues relevant to wireless networks include those dealing with these lower architectural layers, such as data privacy.

Security Threats

The main security issue with wireless networks, especially radio networks, is that they intentionally propagate data over an area that may exceed the limits of the area the organization physically controls. For instance, radio waves easily penetrate building walls and are receivable from the facility's parking lot and possibly a few blocks away. Someone can passively retrieve your company's sensitive information by using the same wireless NIC from this distance without being noticed by network security personnel (see fig. 1.10). This problem also exists with wired ethernet networks, but to a lesser degree. Current flow through the wires emits electromagnetic waves that someone could receive by using sensitive listening equipment. They must be very close to the cable, however, meaning they must first break through physical security.

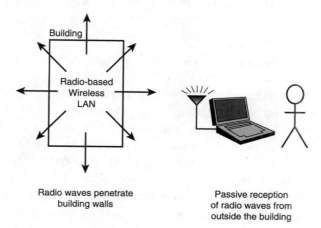

Building

Radio-based
Wireless
LAN

Radio waves penetrate
building walls

Passive reception
of radio waves from
outside the building

Figure 1.10

The passive reception of wireless network data.

Another security problem is the potential for electronic sabotage, in which someone maliciously jams the radio-based network and keeps you from using the network. Remember, wireless networks utilize a carrier sense protocol to share the use of the common medium. If one station is transmitting, all others must wait. Someone can easily jam your network by using a wireless product of the same manufacture that you have within your network and setting up a station to continually resend packets. These transmissions block all stations in that area from transmitting, which is most serious if your company stands to experience a great loss if the network becomes inoperable.

Security Safeguards

Wireless network vendors solve most security problems by restricting access to the data. Most products require you to establish a network access code and set the code within each workstation. A wireless station will not process the data unless its code is set to the same number as the network. Proxim's RangeLAN, for example, can utilize over two billion possible network IDs. If the code is kept secret, it becomes much more difficult for someone to receive and process your data. Some vendors also offer encryption as an option. Lucent's WaveLAN, for example, has two options for encryption. One version encrypts according to the Data Encryption Standard (DES) as defined by the U.S. Department of Commerce, National Institute of Standards and Technology (NIST), formerly called the National Bureau of Standards (NBS). The other version implements a proprietary method called Advanced Encryption Scheme (AES).

The DES and AES algorithms use a 16-digit hexadecimal key for encryption, as shown in figure 1.11. The key is loaded into the security chip when the adapter is configured at installation. When a message is received or sent, the security chip uses the key to encrypt or decrypt the message. Only those workstations in the network with the same security chip and key will be able to understand the messages. Other users of WaveLAN who do not have the key will be unable to decrypt any messages. Both DES and AES perform the encryption in one continuous stream of bits that pass through the system's modulator without affecting performance.

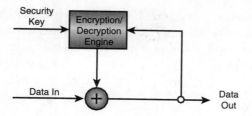

Figure 1.11
The data encryption process.

The Department of Commerce limits export of DES devices outside the U.S. The purpose of the AES is to provide an alternative for DES to those users of WaveLAN needing a secure air interface, but who are not allowed to use DES due to export limitations. AES implements a proprietary algorithm that has been approved for export.

Installation Issues

With wired networks, planning the installation of cabling is fairly straightforward. You can survey the site and look for routes where installers can run the cable. You can measure the distances and quickly determine whether cable runs are possible. If some users are too far away from the network, you can design a remote networking solution or extend the length of the cable by using repeaters. Once the design is complete, installers can run the cables, and the cable plant will most likely support the transmission of data as planned.

A radio-based wireless LAN installation is not as predictable. It is difficult, if not impossible, to design the wireless system by merely inspecting the facility. Predicting the way in which the contour of the building will affect the propagation of radio waves is difficult. Omnidirectional antennas propagate radio waves in all directions if nothing gets in the way. Walls, ceilings, and other obstacles attenuate the signals more in one direction than the other, and even cause some waves to change their paths of transmission. Even the opening of a bathroom door can change the propagation pattern. These events cause the actual radiation pattern to distort, taking on a jagged appearance, as shown in figure 1.12. Wireless MANs are also difficult to plan. What looks like a clear line-of-site path between two buildings separated by 1,500 feet might be cluttered with other radio transmitting devices.

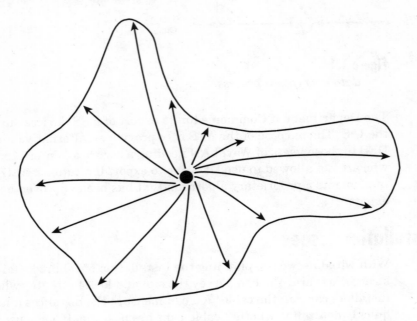

Figure 1.12
The resulting radiation pattern of an omnidirectional antenna within an office building.

To avoid installation problems, an organization should perform propagation tests to assess the coverage of the network. Neglecting to do so may leave some of the users outside of the propagation area of wireless servers and bridges. Propagation tests give you the information necessary to plan wired connections between access points, allowing coverage over applicable areas. Refer to Chapter 8 for identifying the location of access points.

Health Risks

Another common concern is whether wireless networks pose any form of health risk. So far, there has been no conclusive answer. Radio-based networks, however, appear to be just as safe or safer than cellular phones. Many studies have shown little or no risk in using cellular phones, which operate in frequency bands immediately below wireless networks. Wireless network components should be even safer than cellular phones because they operate at lower power levels, typically between 50 and 100 milliwatts, compared to the 600 milliwatts to 3 watt range of cellular phones. In addition, wireless network components usually transmit for shorter periods of time.

Laser-based products, found in both wireless LANs and MANs, offer very little or no health risks. In the U.S., the Center for Devices and Radiological Health (CDRH), a department of the U.S. Food and Drug Administration, evaluates and certifies laser products for public use. The CDRH categorizes lasers into four classes, depending on the amount of harm they can cause to humans. Supermarket scanners and most diffused infrared wireless LANs satisfy Class I requirements, where there is no hazard under any circumstance. Class IV specifies devices, such as laser-scalpels, which can cause grave danger if the operator handles them improperly. Most laser-based wireless MANs rate as Class III devices, whereby someone could damage their eyes if looking directly at the laser beam. Thus, care should be taken when orienting lasers between buildings.

The Wireless Network Market

Wireless networking is applicable to all industries with a need for mobile computer usage or when the installation of physical media is not feasible. Such networking is especially useful when employees must process information on the spot, directly in front of customers. Wireless networking makes it possible to place portable computers in the hands of mobile "front line" workers, such as doctors, nurses, warehouse clerks, inspectors, claims adjusters, real estate agents, and insurance salespeople. The coupling of portable devices with wireless connectivity to a common database and specific applications, as figure 1.13 illustrates, meets mobility needs, eliminates paperwork, decreases errors, and improves efficiency. The alternative to this, which many companies still employ today, is utilizing paperwork to update records, process inventories, and file claims. This manual method processes information slowly, produces redundant data, and is subject to errors caused by illegible handwriting.

The wireless computer approach using a centralized database is clearly superior.

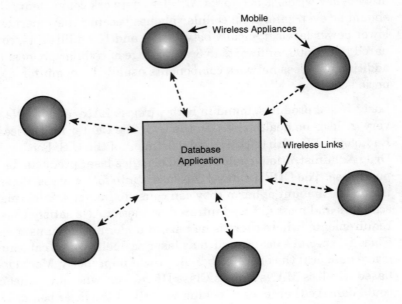

Figure 1.13
A typical configuration for a wireless application.

According to several sources, the mobile wireless LAN market is projected to hit $200 million in sales by the year 1998. Wireless MAN and WAN sales are expected to generate much more. What will motivate people to invest money in wireless networking? This section answers that question by showing the ways in which wireless networks apply to various vertical markets.

N O T E

Refer to Chapters 2–4 for specific case studies describing the applicability of wireless LANs, MANs, and WANs.

Retail

Retail organizations need to order, price, sell, and keep inventories of merchandise. A wireless network in a retail environment enables clerks and storeroom personnel to perform their functions directly from the sales floor. Salespeople are equipped with a pen-based computer or small computing device with bar code reading capability and a wireless link to the store's database. They are then able to complete transactions—such as price checks, special orders, and inventory—from anywhere within the store.

Warehousing

Warehouse staff must manage the receiving, shipping, and inventory of stored goods. These responsibilities keep the staff mobile. Warehouse operations have traditionally been a paper-intensive and time-consuming environment. An organization, however, can eliminate paper, reduce errors, and decrease the time necessary to move items in and out by giving each warehouse employee a handheld computing device with a bar code scanner interfaced via a wireless network to a warehouse inventory system. Upon receiving an item for storage within the warehouse, a clerk can enter the item's nomenclature and part number by keying the information into the database via the handheld device. A forklift operator can then move the item to a storage place and enter the location via a similar handheld device. Thus, the inventory system identifies which items the warehouse contains and where the items are located. As shipping orders enter the warehouse, the inventory system produces a list of the items and their locations. A clerk can view this list from the database via the handheld device and locate the items needed to assemble a shipment. As the clerk removes the items from the storage bins, the database can be updated via the handheld device.

Healthcare

Healthcare centers, such as hospitals and doctors' offices, must maintain accurate records to ensure effective patient care. A simple mistake can cost someone's life. As a result, doctors and nurses must record test results, physical data, pharmaceutical orders, and surgical procedures. This paperwork often overwhelms healthier staff, taking up 50–70 percent of their time. Doctors and nurses are also extremely mobile, going from room to room caring for patients. The use of electronic patient records, with the ability to input, view, and update patient data from anywhere in the hospital, increases the accuracy and speed of healthcare. This improvement is possible by providing each nurse and doctor with a wireless pen-based computer, coupled with a wireless network to databases that store critical medical information about the patients. A doctor caring for someone in the hospital, for example, can place an order for a blood test by keying the request into her handheld computer. The laboratory will receive the order electronically and dispatch a lab technician to draw blood from the patient. The laboratory will run the tests requested by the doctor and enter the results into the patient's electronic medical record. The doctor can then check the results via her handheld appliance from anywhere in the hospital.

Real Estate

Real estate salespeople perform a great deal of their work away from the office, usually talking with customers at the property being sold or rented. Before leaving the office, salespeople normally identify a few sites to show a customer, print the MLS (Multiple Listing Service) information that describes the property, and then drive to each location with the potential buyer. If the customer is unhappy with that round of sites, the real estate agent must drive back to the office and run more listings. Even if the customer decides to purchase the property, they must both go back to the real estate office to finish paperwork that completes the sale. Wireless networking makes the sale of real estate much more efficient. The real estate agent can use a computer away from the office to access a wireless MLS record. IBM's Mobile Networking Group and Software Cooperation of America, for example, have wireless MLS information available that enables real estate agents to access information about properties, such as descriptions, showing instructions, outstanding loans, and pricing. An agent can also use a portable computer and printer to produce contracts and loan applications for signing at the point of sale.

Hospitality

Hospitality establishments check customers in and out and keep track of needs, such as room service orders and laundry requests. Restaurants need to keep track of the names and numbers of people waiting for entry, table status, and drink and food orders. Restaurant staff must perform these activities quickly and accurately to avoid making patrons unhappy. Wireless networking satisfies these needs very well. Someone can greet patrons at the door and enter their name, size of the party, and smoking preference into a common database via a wireless device. The greeter can then query the database and determine the availability of an appropriate table. Those who oversee the tables also would have a wireless device used to update the database to show whether the table is occupied, being cleaned, or available. After obtaining a table, the waiter will transmit the order to the kitchen via the wireless device, eliminating the need for paper order tickets.

Utilities

Utility companies operate and maintain a highly distributed system that delivers power and natural gas to industries and residences. Utility companies must continually monitor the operation of the electrical distribution system and gas lines, and must check usage meters at least monthly to calculate bills. Traditionally, this means a person must travel from location to location, record information, and then enter the data at a service or computing center. Several utility companies are employing wireless networks to support the automation of meter reading and system monitoring, saving time and reducing overhead costs. Kansas City Power & Light, for example, operates one of the largest wireless metering systems, serving more than 150,000 customers in eastern Kansas and western Missouri. This system employs a monitoring device at each customer site that takes periodic meter readings and sends the information back to a database that tracks usage levels and calculates bills, avoiding the need for a staff of meter readers. In addition, the Jacksonville Electric Authority (JEA) in Jacksonville, Florida, uses a RAM Mobile Data wireless WAN service to save time and reduce paperwork. This system eliminates radio conversations and paperwork between central-site dispatchers and maintenance people, speeding up the service to customers.

Field Service

Field service personnel spend most of their time on the road installing and maintaining systems or inspecting facilities under construction. In order to complete their jobs, these individuals need access to product documentation and procedures. Traditionally, field service employees have had to carry several binders of documentation with them to sites that often lack a phone and even electricity. In some cases, the field person might not be able to take all the documents with him to a job site, causing him to delay the work while obtaining the proper information. On long trips this information may also become outdated. Updates require delivery that may take days to reach the person in the field. Wireless access to documentation can definitely enhance field service. A field service employee, for example, can carry a portable computer connected via a wireless network to the office LAN containing accurate documentation of all applicable information.

Field Sales

Sales professionals are always on the move meeting with customers. While on site with a customer, a salesperson needs access to vast information that describes products and services. Salespeople must also place orders, provide status, such as meeting schedules, to the home office, and maintain inventories. With wireless access to the home office network, a salesperson can view centralized contact information, retrieve product information, produce proposals, create contracts, and stay in touch with home office staff and other salespeople. This contact permits salespeople to complete the entire sale directly from the customer site, which increases the potential for a successful sale and shortens the sales cycle.

Vending

Beverage and snack companies place vending machines in hotels, airports, and office buildings to enhance the sales of their products. Vending machines eliminate the need for a human salesclerk. These companies, however, must send employees around to stock the machines periodically. In some cases, machines might become empty before the restocking occurs because the company has no way of knowing if the machine runs out of a particular product. An empty machine does not generate revenue and is therefore of no value to the company. A wireless network, though, can support the monitoring of stock levels by transporting applicable data from each of the vending machines to a central database that can be easily viewed by company personnel from a single location. Such monitoring allows companies to be proactive in stocking their machines because they will always know the stock levels at each machine. Comverse Technology's DGM&S subsidiary licensed software to BellSouth to support a vending machine monitoring service called Cellemetry, which uses the data channels of existing cellular networks.

The Future of Wireless Networks

Where is wireless networking going? What will the future bring? Predicting what the state of this technology and its products will be five years from now, or even a year from now, is impossible. The outlook for wireless networks, however, is very good. As figure 1.14 illustrates, the maturation of standards should motivate vendors to produce new wireless products and drive the prices down to levels that are much easier to justify. The presence of standards will motivate smaller companies to manufacture wireless components because they will not need to invest large sums of money in the research and development phases of the product. These investments will have already been made and embodied within the standards, which will be available to anyone interested in building wireless network components.

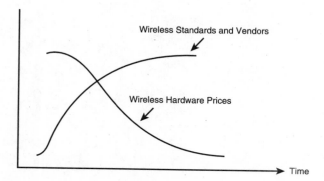

Figure 1.14
The future of wireless networking.

Wireless Local Area Networks (LANs)

Do you need wire-free interconnectivity between users' computers, servers, and printers within a room or building? A wireless LAN might be a feasible alternative to traditional ethernet and token-ring networks to satisfy needs for user mobility in offices, warehouses, retail stores, and hospitals.

This chapter describes several approaches to wireless networking within a local environment. These methods include the following:

○ Radio waves

○ Infrared light

○ Carrier currents

Wireless LAN Applications

Figure 2.1 illustrates the concept of a wireless LAN. Most wireless LANs operate over unlicensed frequencies at near-ethernet speeds (10 Mbps) using carrier sense protocols to share a radio wave or infrared light medium. The majority of these devices are capable of transmitting information up to 1,000 feet between computers within an open environment, and their costs per user range from $150–$800. In addition, most wireless LAN products offer Simple Network Management Protocol (SNMP) to support network management through the use of SNMP-based management platforms and applications.

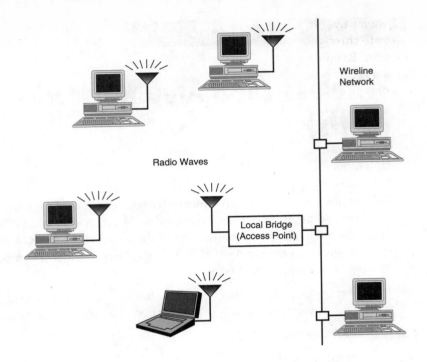

Figure 2.1
A wireless local area network.

Wireless LANs can save money when installing networks in difficult-to-wire environments, such as old facilities containing asbestos, historical buildings, and jails. The removal of asbestos can cost thousands of dollars to facilitate the installation of wiring. Historical societies forbid changes, such as holes and wallplates, in certain building structures. Jails consist of large amounts of concrete and steel to keep prisoners from escaping. As a result, justice departments do not allow the drilling of holes for cable runs because it might weaken security. In these cases, as well as others where it's expensive or time consuming to install wire, consider using a wireless LAN.

Radio-based Wireless LANs

The most widely sold wireless LAN products use radio waves as a medium between computers and peripherals. An advantage of radio waves over other forms of wireless connectivity is that they propagate through walls and other obstructions with fairly little attenuation. Even though several walls might separate the user from the server or wireless bridge, users can maintain connections to the network—supporting true mobility. With radio-LAN products, a user with a portable computer can move freely through the facility while accessing data from a server or running an application.

A disadvantage of using radio waves, however, is that an organization must manage the radio waves along with other electromagnetic propagation. Medical equipment and industrial components utilize the same radio frequencies as wireless LANs, which could cause interference. As mentioned in Chapter 1, "Introduction to Wireless Networking," an organization must determine whether potential interference is present before installing a radio-based LAN. Because radio waves penetrate walls, security may also be a problem. Unauthorized people from outside the controlled areas could receive sensitive information. As mentioned in Chapter 1, however, vendors often scramble the data signal to protect the information from being understood by inappropriate people.

This section discusses the following radio-based wireless LAN topics:

- ○ ISM bands
- ○ Narrow band wireless LANs
- ○ Spread spectrum wireless LANs

ISM Bands

In 1985, as an attempt to stimulate the production and use of wireless network products, the FCC modified Part 15 of the radio spectrum regulation, which governs unlicensed devices. The modification authorized wireless network products to operate in the Industrial, Scientific, and Medical (ISM) bands. The ISM frequencies are shown in figure 2.2. The FCC allows users to operate wireless products without obtaining FCC licenses if the products meet certain requirements, such as operation under 1 watt transmitter output power. This deregulation of the frequency spectrum eliminates the need for user organizations to perform costly and time-consuming frequency planning to coordinate radio installations that will avoid interference with existing radio systems. This is even more advantageous if you plan to move your equipment frequently because you can avoid the paperwork involved in relicensing the product at the new location. As you can see, more bandwidth is available within the higher frequency bands, which will support higher data rates.

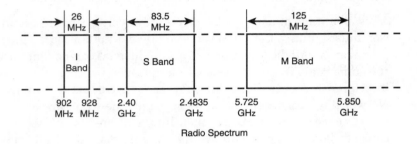

Figure 2.2
The Industrial, Scientific, and Medical (ISM) frequency bands.

ISM Band Availability

The ISM band frequencies are not available in all parts of the world, limiting the capability to operate wireless products sold in the United States. Figure 2.3 identifies those countries that allow wireless LAN operation in the 902 MHz and 2.4 GHz ISM bands. The 2.4 GHz is the only unlicensed band available worldwide. This band was approved in North and South America in the mid-1980s and was accepted in Europe and Asia in 1995. Companies first began developing products in the 902 MHz band because manufacturing costs in this band were cheaper. The lack of availability of this band in some areas and the need for greater bandwidth, however, drove these companies to migrate many of their products to the 2.4 GHz band.

Narrow Band Modulation

Conventional radio systems, such as television and AM/FM radio, utilize narrow band modulation. These systems concentrate all their transmit power within a narrow range of frequencies, making efficient use of the radio spectrum in terms of frequency space. The idea behind most communications design is to conserve as much bandwidth as possible; therefore, most transmitted signals utilize a relatively narrow slice of the radio frequency spectrum. Other systems using the same transmit frequency, however, will cause a great deal of interference because the noise source will corrupt most of the signal. To avoid interference, the FCC requires users of narrow band systems to obtain FCC licenses to properly coordinate the operation of radios. Narrow band products thus have a strong advantage because you can be fairly assured of operating without interference. If interference does occur, the FCC will resolve the matter. This makes narrow band modulation good for longer links covering the geographical size of a metropolitan area.

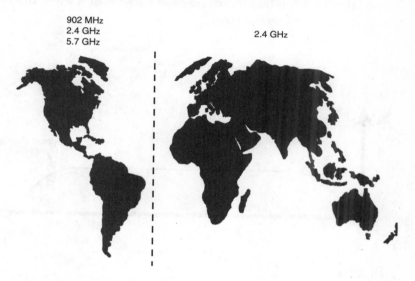

Figure 2.3
ISM spectrum availability.

N O T E

Chapter 3, "Wireless Metropolitan Area Networks," covers the wireless MAN products that employ narrow band radio frequencies.

Spread Spectrum Modulation

Products that operate according to Part 15.247 of the FCC's Rules and Regulations must utilize spread spectrum modulation. What is spread spectrum? Spread spectrum modulation "spreads" a signal's power over a wider band of frequencies (see fig. 2.4). This contradicts the desire to conserve frequency bandwidth, but the spreading process makes the data signal much less susceptible to electrical noise than conventional radio modulation techniques. Other transmission and electrical noise, typically narrow in bandwidth, will only interfere with a small portion of the spread spectrum signal, resulting in much less interference and less errors when the receiver demodulates the signal.

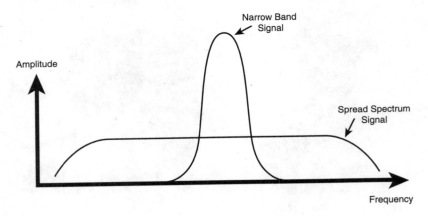

Figure 2.4
Narrow band versus spread spectrum modulation.

> **N O T E**
>
> Spread spectrum was initially developed by the U.S. military during World War II to protect communication systems and guided weapons from intentional hostile jamming. One of the principle developers of spread spectrum was Hedy Lamarr, an actress during the 1940s. Hedy had invented the modulation technique to prevent the enemy from jamming or eavesdropping on secret military conversations. One of her first devices was to keep guided torpedoes from being detected or jammed by the enemy. Hedy and George Antheil, a film-score composer who assisted Hedy in perfecting spread spectrum, received a patent for their work in 1940. Actually, spread spectrum

was never used during World War II. Sylvania utilized spread spectrum for the first time on ships sent to blockade Cuba in 1962. Hedy Lamarr conceived an excellent modulation technique; however, she never received any compensation for the idea.

Spread spectrum modulators use one of two methods to spread the signal over a wider area: direct sequence or frequency hopping.

Direct Sequence Spread Spectrum

Direct sequence spread spectrum combines a data signal at the sending station with a higher data rate bit sequence, which many refer to as a *chipping code* (also known as *processing gain*). A high processing gain increases the signal's resistance to interference. The minimum linear processing gain that the FCC allows is 10, and most products operate under 20. The IEEE 802.11 Working Group has set their minimum processing gain requirements at 11.

Figure 2.5 shows an example of the operation of direct sequence spread spectrum. A chipping code is assigned to represent logic "one" and "zero" data bits. As the data stream is transmitted, the corresponding code is sent. For example, the transmission of a data bit equal to "one" would result in sequence 00010011100 being sent.

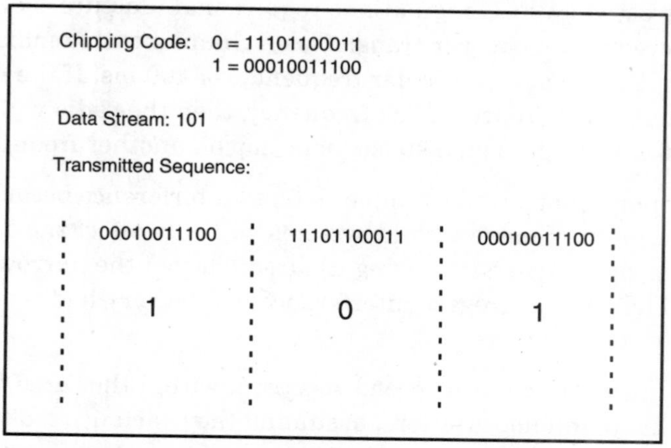

```
Chipping Code:   0 = 11101100011
                 1 = 00010011100

Data Stream: 101

Transmitted Sequence:

   00010011100    :    11101100011    :    00010011100

        1         :         0         :         1
```

Figure 2.5
The operation of direct sequence spread spectrum.

Many direct sequence products on the market utilize more than one channel in the same area; the number of channels available, however, is limited. With direct sequence, many products operate on separate channels by slicing the frequency band into non-overlapping frequency channels. This results in the potential for several separate networks to operate without interfering with each other. To leave enough bandwidth for moderate to high data rates, however, there can only be a few channels. Proxim's ProxLink and RangeLAN product families, for example, use direct sequence technology in the 902–928 MHz frequency band. ProxLink incorporates seven different channels, and RangeLAN uses three channels.

Frequency Hopping Spread Spectrum

Frequency hopping works very much like its name implies. It takes the data signal and modulates it with a carrier signal that hops from frequency-to-frequency as a function of time over a wide band of frequencies (see fig. 2.6). A frequency hopping radio, for example, will hop the carrier frequency over the 2.4 GHz frequency band between 2.4 GHz and 2.483 GHz. A hopping code determines the frequencies the radio will transmit and in which order. To properly receive the signal, the receiver must be set to the same hopping code and "listen" to the incoming signal at the right time and correct frequency. FCC regulations require manufacturers to use 75 or more frequencies per transmission channel with a maximum dwell time (time at a particular frequency) of 400 ms. If the radio encounters interference on one frequency, then the radio will retransmit the signal on a subsequent hop on another frequency.

The frequency hopping technique reduces interference because the propagation from narrow band systems will only affect the spread spectrum signal when it is using the frequency of the narrow band signal. Thus, the aggregate interference will be very low, resulting in little or no bit errors.

Operating radios can use spread spectrum within the same frequency band and not interfere, assuming they each use a different hopping pattern. While one radio is transmitting at one particular frequency, the other radio(s) uses a different frequency. A set of hopping codes that never use the same frequencies at the same time are considered *orthogonal*. Some vendors allow the user to choose the channel (a particular hopping code) through software that the radio will operate on, all users within the same local network, however, have to use the same code. This does give you the ability, though, to have wireless LANs within close proximity to each other operate within the same band and not interfere with each other, as long as you assign them orthogonal hopping codes. The FCC's requirement for the number of different transmission frequencies allows frequency-hopping radios to have many non-interfering channels.

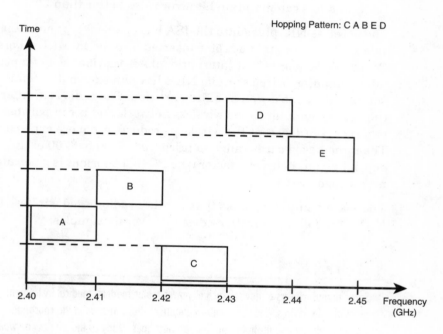

Figure 2.6
A frequency hopping spread spectrum.

Radio-based Wireless LAN Components

The components of a radio-based LAN consist mainly of a wireless NIC and a wireless local bridge, which is sometimes referred to as an *access point*.

Wireless NICs

The wireless NIC interfaces the computer to the wireless network by modulating the data signal with the spreading sequence and implementing a carrier sense access protocol. If the computer needs to send data onto the network, the NIC will listen for other transmissions. If the NIC doesn't sense other transmissions, it will transmit a frame of data. Other stations, constantly listening for incoming data, capture the transmitted frame and check whether its address matches the destination address in the frame's header. If there is a match, the receiving station will process the frame. If not, the station will discard the frame. Most radio-based wireless LAN products can maintain bit error rates better than 10^{-8}.

The wireless NIC plugs into the ISA bus of a desktop computer or a small credit card-sized adapter inserted into the PCMCIA (Personal Computer Memory Card International Association) slot of a portable computer. Wireless radio NICs have an external antenna that you can attach to a wall or office partition. The radio NIC interfaces the users' computer to the wireless network and is comparable to an ethernet board, except the wireless products are more expensive. The radio-based cards range in price from $300 to $700 each, depending on their form factor (the PCMCIA format is generally most expensive).

There are many companies that produce radio-based wireless LAN NICs. The following sections discuss several examples:

N O T E

Some manufacturers of devices such as power meter readers, medical instruments, point-of-sale devices, and order entry equipment, have interest in incorporating wireless data communications into their product lines. Many of the wireless network companies can support this need by offering Original Equipment Manufacture (OEM) versions of their products. Proxim, for example, has RangeLAN2 6300 Mini-ISA that measures under four inches long. RangeLAN2 6300 is an integrated hardware/software OEM package that enables developers to easily incorporate wireless networking capabilities into their products. It is especially well suited for integration into portable computer platforms for mobile data applications. Another company, Digital Wireless Corporation, offers an OEM product called WIT2400 Frequency Hopping Transceiver. The WIT2400 operates over the 2.4 GHz ISM band at up to 2000 feet with data rates of 250 Kbps.

Proxim's RangeLAN

Proxim, Inc. has a full family of radio-based wireless LAN products. In 1993, Proxim became the first company to ship a spread spectrum PCMCIA wireless LAN adapter, the RangeLAN2 7200 PCMCIA. The RangeLAN2 7200 is a high-performance wireless LAN adapter for PCMCIA Type II or III equipped portables. RangeLAN2 7200 uses frequency-hopping spread spectrum radio signals in the 2.4–2.4835 GHz band, delivering a data rate of 1.6 Mbps at up to 3,000 feet. RangeLAN2 7200 operates at a distance of up to 500 feet in normal office environments and up to 1000 feet in open spaces. Through a multi-channel approach, RangeLAN2 7200 enables 15 independent wireless LANs to operate within the same physical space. Proxim also sells RangeLAN2 7100 ISA for AT bus computers, and it has the same performance characteristics as the PCMCIA version.

Lucent's WaveLAN

Lucent has several versions of their wireless LAN product WaveLAN. You can purchase WaveLAN for ISA or PCMCIA formats at either 902 MHz or 2.4 GHz. WaveLAN's range in open environments is 600–800 feet for the 902 MHz version and 12–180 feet for the 2.4 MHz model. The entire WaveLAN family uses direct sequence spread spectrum. For higher levels of security, WaveLAN will support data encryption. This product has undergone several ownership changes. The initial product was developed by NCR, which was acquired by AT&T and renamed AT&T Global Information Solutions. AT&T then split into three parts: the service groups still belong to AT&T, the computer section is with NCR again, and WaveLAN is part of Lucent Technologies.

Windata's FreePort

Windata's FreePort Wireless Ethernet system consists of two main hardware components: FreePort Wireless Hub and Wireless Transceiver. The system operates at 5.7 Mbps throughput and has a range of up to 260 feet. The FreePort Wireless Hub receives and retransmits data packets from Wireless Transceivers in its coverage area. The Wireless Hub can interface directly to IEEE 802.3 ethernet, acting as a wireless bridge.

AeroComm's GoPrint Wireless Printer Sharing

GoPrint features a 2.4 GHz spread spectrum radio that enables a computer to interface with a printer at distances of 3,000 feet unobstructed and up to 200–500 feet where office walls are present. GoPrint transmits at speeds up to 1 Mbps, which is 5–10 times the data throughput of wired printer sharing solutions. GoPrint adapters plug directly into the parallel port of computers and the Centronics port of printers. Multiple computers can output to a single printer or up to eight printers in the same office. GoPrint will work in both server-oriented and non-server environments.

Wireless Local Bridges

Network bridges are an important part of any network—they interconnect multiple LANs at the Medium Access Control (MAC) layer to produce a single logical network. The MAC layer, which provides medium access functions, is part of IEEE's architecture describing LANs. Chapter 8, "Designing a Wireless Network," describes this architecture. Bridges interface LANs together, such as ethernet-to-ethernet or ethernet-to-token ring, and also provide a filtering of packets based on their MAC layer address. This allows an organization to create segments within an enterprise network. If a networked station sends a packet to another station located on the same segment, the bridge will not forward the packet to other segments or the enterprise backbone. If the packet's destination is on a different segment, however, the bridge will allow the packet to pass through to the destination segment. Thus, bridges ensure that packets do not wander into parts of the network where they are not needed. This process, known as *segmentation*, makes better use of network bandwidth and increases overall performance.

There are two types of bridges—local and remote. *Local bridges* connect LANs within close proximity, and *remote bridges* interconnect sites that are separated by distances greater than the LAN protocols can support. Figure 2.7 illustrates the differences between local and remote network bridges. Traditionally, organizations have used leased digital circuits, such as T1 and 56 Kbps, to facilitate the connections between a pair of remote bridges. Chapter 3, "Wireless Metropolitan Area Networks," describes wireless remote network bridging that uses radio and infrared light as the medium.

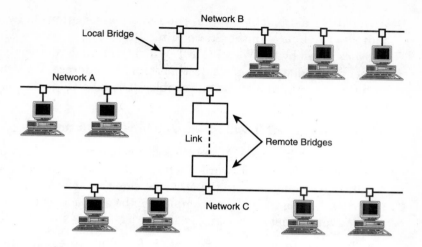

Figure 2.7
Local versus remote bridges.

Most companies that build wireless LAN NICs also sell a wireless local bridge. Proxim, for example, sells a bridge called RangeLAN2 7500 Access Point that interfaces the wireless RangeLAN2 family products to an IEEE 802.3 ethernet network. RangeLAN2 7500 operates in the 2.4–2.4835 GHz ISM band using frequency-hopping spread spectrum. RangeLAN2 7500 optimizes the network's performance and reliability by filtering local packets and only forwarding packets meant for other network segments. RangeLAN2 7500 automatically learns source addresses by monitoring network traffic. Once identified, the address information is stored and forwarded within the network, resulting in an overall reduction in network traffic. RangeLAN2 7500 enables roaming, allowing mobile computing users to move seamlessly from one RangeLAN2 wired access point to another without losing their network connection.

The filtering process of a local bridge is critical in maintaining a network configuration that minimizes unnecessary data traffic. WavePOINT, a wireless local network bridge that is part of the WaveLAN product family, has a filter table that contains MAC layer addresses mapped to either the WaveLAN or ethernet side of the bridge. When the bridge receives a packet, the bridge creates a record containing the MAC address and the port it receives the frame on in a dynamic table.

WavePOINT also enables you to enter static associations between addresses and ports in the Static Filter Table. These entries cannot be overwritten by the WavePOINT. When the bridge receives a frame, the bridge looks at its destination MAC address, then checks both the dynamic and static filter tables. The following situations may occur:

1. The WavePOINT will forward all broadcast frames.

2. If the destination MAC address is not in either filter table, the WavePOINT will forward the frame to the opposite port. A frame coming in from the ethernet side that does not have an entry in the tables, for example, will be sent across to the WaveLAN network segment.

3. If the destination MAC address is found in either filter table, the WavePOINT will decide whether to forward the frame based on what it finds in the table. The WavePOINT, for example, will not forward a frame coming in from the ethernet side and having an association corresponding to the ethernet side. It will forward the frame, however, if it had an association with the WaveLAN side.

Radio-based Wireless LAN Configurations

The combination of wireless NICs and bridges gives network managers and engineers the ability to create a variety of network configurations. There are two main configurations a wireless LAN can assume—single cell and multiple cell.

Single-Cell Wireless LANs

For small single-floor offices or factories, a single-cell wireless LAN covering roughly a million square feet may suffice. Single-cell wireless LANs only require wireless NICs in devices connected to the network, as shown in figure 2.8. Access points are not necessary. For example, Xircom's Netwave wireless LAN product, as with most other wireless LAN products, allows several configurations. You can create a spontaneous LAN easily using Netwave equipped portables without the use of any access points to form a peer-to-peer network. Any time two or more PCMCIA adapters are within range of each other, they can establish a peer-to-peer network. This allows an organization to form an ad hoc network for temporary use.

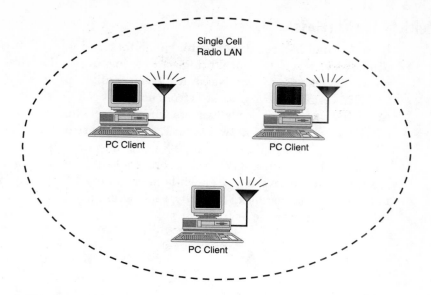

Figure 2.8
A single-cell wireless LAN.

With Xircom's Netwave products, the area covered by stations within a peer-to-peer network is called a *Basic Service Area* (BSA). A BSA covers approximately 150 feet between all units in a typical office environment (650 feet in open areas). A single radio-based wireless LAN segment, such as the BSA, can support 6–25 users and still keep network access delays at an acceptable level. These networks require no administration or preconfiguration.

With Netwave, there are two types of peer-to-peer networks— public and named. *Public networks* use a default domain, allowing any Netwave user within range to join the network. If you need more privacy, you can create a *named network* by having users configure the Netwave adapters with a specific Domain ID. With this configuration, only stations having the same Domain ID can join the network. The optional data scrambling feature requires participants to know the scrambling key used by the network to decode data packets from other stations on the network. In addition, the Domain ID specifies a unique hopping code that minimizes interference between adjacent wireless networks. Most vendors utilize similar approaches.

Multiple Cell Wireless LANs

If an organization requires greater range than a single cell, it can utilize a set of wireless local bridges (access points) and a wired network backbone to create a multiple-cell configuration (see fig. 2.9). This enables wireless users from different cells to communicate with each other, as well as lets wireless users access resources available on the wired network. Such a configuration can cover larger multi-floor buildings, campuses, and hospitals. In this environment, a portable PC with a cordless LAN adapter can also roam within the coverage area while maintaining a live connection to the corporate network. Typically, each access point utilizes a different hopping code or frequency.

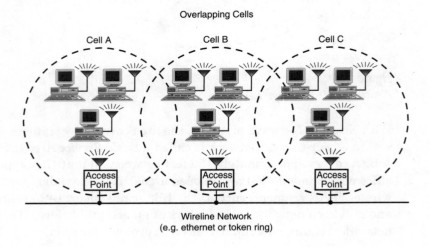

Figure 2.9
A multiple-cell wireless LAN.

The ideal wireless LAN configuration for your organization depends primarily on user requirements and geography. If you have a relatively small group that requires wireless interconnectivity within the immediate group, a single cell may do the job. If users are spread throughout the entire facility, however, then you might need a multiple-cell configuration. In either case, bridges may be necessary to support user access to resources located on the wired infrastructure.

A critical function in a multiple-cell network is *roaming*, which allows wireless users to move from cell to cell seamlessly. Most wireless LAN companies who build wireless network bridges implement roaming. The roaming protocol works only at the MAC layer; therefore, it will not work over routers.

This paragraph explains how WaveAROUND, which belongs to the WaveLAN product line, implements roaming[1]. All cells in a WaveLAN configuration are linked together in a wireless network group, called the Domain. Within this Domain, a mobile station will automatically switch between different cells to ensure continuous connectivity. The mobile station will monitor the communications quality with the WavePOINT of the current cell. If the communications quality drops below a preset value, the station will start searching for another cell. If found, the station will retrieve and adopt the WaveLAN Network ID of a new cell to ensure the network connection.

With WaveAROUND, multiple cells do not have to overlap. If a wireless user moves between two non-overlapping cells, as shown in figure 2.10, the WaveAROUND automatically reconnects the user to the new cell upon entering the new location. In situations where cells do overlap, WaveAROUND connects to a new WavePOINT before terminating the connection with the predecessor. This provides a constant connection to the network.

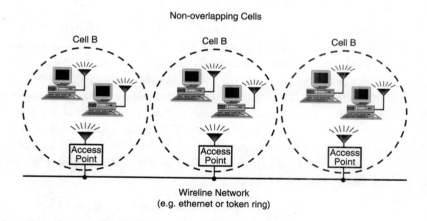

Figure 2.10
A multiple cell wireless network with non-overlapping cells.

How does roaming work? Again, let's use WavePOINT as an example. The WavePOINT broadcasts beacon messages at regular intervals to support roaming mobile stations. Beacon messages contain the Domain ID, the WaveLAN Network ID of the WavePOINT, communications quality information, and cell search threshold values. The Domain ID identifies the WavePOINTs and mobile stations that belong to the same WaveLAN roaming network. A mobile station listening for beacons will only interpret beacon messages with the same Domain ID. The WaveLAN Network ID identifies a specific cell of the WaveLAN network. This is the Network ID assigned to each WavePOINT. It tells the mobile station which WaveLAN Network ID to use to communicate with the network on that location. You can encode the Network ID information by using a "beacon key." Communications quality information helps the mobile station determine the quality of the link to the WavePOINT of a particular cell. The cell search thresholds are related to the level of communications quality. They activate the cell search mode when the communications quality drops, such as when the mobile station is moving to another location. In cell search mode, the mobile station will start searching for other WavePOINTs by listening for beacons. You can change the threshold values by selecting a different level of sensitivity, depending on your environment.

A roaming station needs a number of beacons to determine the communications quality with a WavePOINT. Responsiveness parameters set beacon interval time and beacon time-out. The beacon interval time allows you to set the frequency of beacon transmissions. This frequency determines how fast a station can decide the actual communications quality. The beacon time-out sets the values of a timer mechanism. This timer activates the cell search mode when the mobile station does not receive a beacon within the specified time limit. Three user defined presets adjust the level of responsiveness:

○ **Relaxed.** If responsiveness is Relaxed, the beacon frequency will be low. This setting avoids unnecessary use of the mobile stations' processing capacity if needed to interpret beacon messages.

○ **Normal.** If responsiveness is Normal, the beacon frequency will be moderate.

○ **Fast.** If responsiveness is Fast, the beacon frequency will be high. A mobile station will be able to react faster to a deterioration of communications quality, avoiding communication errors. Faster responsiveness is necessary for interactive mobile applications, whereas stationary operations can function with slow (relaxed) response.

Sensitivity determines the amount of time the mobile station will spend in cell search mode and when the mobile station will switch to another WavePOINT. Sensitivity parameters set the values of the Cell Search Thresholds, which determine when the mobile station starts or stops looking for another WavePOINT. These thresholds are related to the level of communications quality. There are three Cell Search thresholds:

○ **Regular Cell Search.** The level of communications quality at which the mobile station starts looking for another WavePOINT. The station will only switch over to a new WavePOINT if the level of communications quality with that WavePOINT is higher than the Stop Cell Search threshold.

○ **Fast Cell Search.** The level of communications quality at which the mobile station starts looking for a WavePOINT with any acceptable level of communications quality. In this case, the station will immediately switch over to a WavePOINT that provides better communications quality.

○ **Stop Cell Search.** The level of communications quality at which the mobile station stops looking for a WavePOINT. There are three user-defined Sensitivity presets: Low, Normal and High.

○ **Low.** If sensitivity is Low, a roaming station will stay connected to a WavePOINT as long as possible. It will start searching for another WavePOINT later than with the Normal setting and stop searching earlier. The Low sensitivity settings are best when coverage areas are not adjacent to one another. They will avoid a station looking for WavePOINTs when there is no WavePOINT within range.

○ **Normal.** If sensitivity is Normal, a roaming station will work best in most environments.

○ **High.** If sensitivity is High, a roaming station will try to switch to another WavePOINT as soon as possible. It will start searching for another WavePOINT earlier and stop searching later than with the Normal setting. If Sensitivity is High, for example, a roaming station is likely to spend more time in cell search mode.

In cell search mode, a mobile station has to interpret beacons and network broadcast messages transmitted by different WavePOINTs. If sensitivity is too high, a station is likely to spend more time in cell search mode than needed. This will cause unnecessary use of the processing capacity from the mobile station. A mobile station requires network overhead to switch between two WavePOINTs.

WaveAROUND roaming functionality enables a mobile station to detect an out-of-range situation and reestablish a lost connection. The combined characteristics of applications and the NOS, however, may pose problems for network operations because most of today's applications are not designed for use in a wireless mobile environment. Future developments of applications should allow for temporarily working offline. When a connection is lost, the application must be able to "synchronize files" as soon as the connection becomes available again.

Radio-based Wireless LAN Performance

Radio-based wireless LANs offer performance similar to ethernet networks. Figure 2.11 compares the performance of WaveLAN versus ethernet. The figure shows the response time of performing a DOS file copy for several different size files between a 80386/25 MHz server and 80386sx/16 MHz workstation via WaveLAN NICs. For file sizes of less than 100 KB, ethernet and WaveLAN performance is nearly the same. For larger files, though, ethernet takes the lead. The actual performance will depend on the application's file sizes and frequency of network use.

In addition, radio-based wireless bridges were designed to operate within a typical LAN environment. WavePOINT, for example, was designed to operate under the following assumptions:

○ Ethernet network utilization = 20%

○ Frame size varying from 64–1518 bytes

○ WaveLAN throughput of 150 KByte/s

○ Traffic to be forwarded = 25%

This allows the bridge to keep up with typical ethernet traffic.

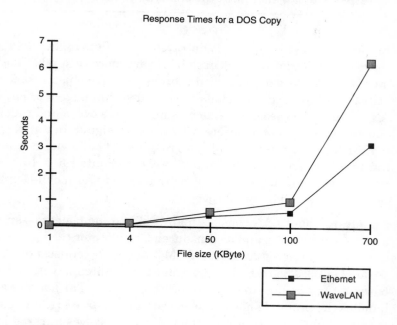

Figure 2.11
WaveLAN versus 10 Mbps ethernet.

Infrared Light-based Wireless LANs

Infrared light is an alternative to using radio waves for wireless
LAN interconnectivity. The wavelength of infrared light ranges
from about 0.75 to 1,000 microns, which is longer (lower in frequency) than the spectral colors but much shorter (higher in
frequency) than radio waves. Under most lighting conditions,
therefore, infrared light is invisible to the naked eye. Infrared
light LAN products operate around 820 nanometer wavelengths
because air offers the least attenuation at that point in the infrared
spectrum.

> **NOTE**
>
> Sir William Herschel discovered infrared light in 1800 when he separated sunlight into its component colors with a prism. He found that most of the heat in the beam fell in the spectral region where no visible light existed, just beyond the red.

In comparison to radio waves, infrared light offers higher degrees of security and performance. These LANs are more secure because infrared light does not propagate through opaque objects, such as walls, keeping the data signals contained within a room or building. Also, common noise sources such as microwave ovens and radio transmitters will not interfere with the light signal. In terms of performance, infrared light has a great deal of bandwidth, making infrared light possible to operate at very high data rates. Infrared light, however, is not as suitable as radio waves for mobile applications because of its limited coverage.

An infrared light LAN consists mainly of two components—an adapter card or unit and a transducer. The adapter card plugs into the PC or printer via an ISA or PCMCIA slot (or connects to the parallel port). The transducer, similar to the antenna with a radio-based LAN, attaches to a wall or office partition. The adapter card handles the protocols needed to operate in a shared medium environment, and the transducer transmits and receives infrared light signals.

There are two types of infrared light LANs:

○ Diffused

○ Point-to-point

Diffused Infrared-based LAN Technique

You've probably been using a diffused infrared device for years—the television remote control—which allows you to operate your TV from a distance without the use of wires. When you depress a button on the remote, a corresponding code modulates an infrared light signal that is transmitted to the TV. The TV receives the code and performs the applicable function. This is fairly simple, but infrared-based LANs are not much more complex. The main difference is that LANs utilize infrared light at slightly higher power levels and use communications protocols to transport data.

When using infrared light in a LAN, the ceiling can be a reflection point (see fig. 2.12). This technique uses carrier sense protocols to share access to the ceiling. Imagine, for example, a room containing four people who can only communicate via flashlights. To send information, the people can encode letters that spell words using a system such as Morse Code. If someone wants to send information, they first look at the ceiling to see if someone is currently transmitting (shining light off the ceiling). If a transmission is taking place, the person wanting to send the information waits until the other person stops sending the message. If no one is transmitting, the source person will point their flashlight to the ceiling and turn the light on and off, according to the code that represents the information being sent. To alert the destination person of an incoming message, the sender will transmit the proper sequence of code words that represent the destination person's name. All people in the room will be constantly looking at the ceiling, waiting for light signals containing their address. If a person "sees" their name, they will pay attention to the rest of the transmission. Through this method, each person will be able to send and receive information.

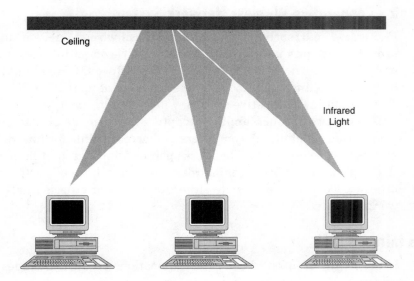

Figure 2.12
A diffused infrared-based wireless LAN technique.

Diffused infrared light LANs work similarly to the previous analogy. The LANs, however, operate much faster. Typical data rates are 1–4 Mbps versus the 20–30 bps the average person can send using flashlights and Morse Code. Prices for these types of wireless adapters range from $200 to $500 each.

Due to geometry, diffused infrared light stations are limited in separation distance, typically 30–50 feet. The lower the ceiling, the less range between stations. Ceiling heights of ten feet will limit the range to around forty feet. To extend the operating range, you can utilize infrared access points to connect cells together via a wired backbone.

NOTE

Because they depend on ceilings and walls, diffused infrared light LANs will not operate outdoors.

Several companies sell diffused infrared LAN components. The following sections provide an overview of these products.

Photonic's Cooperative Wireless Network System

Photonics is currently the only dual platform wireless networking provider with products for both Apple and PC-compatible computers. Photonics' wireless LAN products use diffuse infrared light that bounces off ceilings to provide a coverage area of up to 25' × 25'. The Photonics Cooperative system, which operates at 1 Mbps, consists of two product families: adapters that provide peer-to-peer networking capability to computers and access points for linking groups of wireless users. The access points, which support multiple cell configurations by providing roaming from cell to cell, will interface Cooperative users to either AppleTalk or ethernet networks.

IBM's Infrared LAN

IBM's family of infrared LAN communications products includes infrared adapter models that have form factors fitting Micro Channel, ISA, and PCMCIA Type II.

IBM offers two types of Access Points. The Access Point Unit gives multiple mobile computing users and wireless cells access to ethernet LANs. It also enables multiple wireless cells to communicate with each other via a wired LAN. The Access Point Unit comes with a built-in ethernet and a PCMCIA Infrared LAN adapter with a tethered transceiver. The Access Point Software enables a 33MHz 486 (or higher) PC with an ethernet or token-ring adapter to act as an infrared LAN access point. This software comes at no charge with the purchase of an IBM Infrared LAN adapter.

Spectrix's SpectrixLite

Spectrix produces a diffused infrared LAN called SpectrixLite for portable computers. A SpectixLite Communications Link connects to a PC in a variety of formats: RS-232, TTL, RS-485/RS-422, or PCMCIA Type II. The Comm Link sends and receives diffused infrared light with a wide angle field of view and offers a 4 Mbps data rate. With the signal bouncing off walls and objects, communications are omnidirectional, and the users can move freely within a 50 foot range of the Base Station. The Base Station provides wireless subnet control using Spectrix's patented deterministic reservation/polling wireless protocol called CODIAC (Centralized Operation Deterministic Interface Access Control). CODIAC minimizes consumption of a portable's battery power and supports time-bounded services by making stations that want to transmit information reserve time slots to send their data. Antennas are connected to the Base Station by twisted-pair wires, providing a 40,000 square foot service area using a 16-port hub. The system enables users to roam seamlessly between Base Stations while maintaining connectivity.

Point-to-Point Infrared Techniques

Some infrared products operate in a point-to-point manner, that is, where the devices maintain direct links with one another. Two highly different products in this category are the "point and beam" devices that transfer files directly between computers and peripherals, and a point-to-point infrared LAN system that replaces the wire in a token-ring network with infrared light.

"Point and Beam" Infrared Links

"Point and beam" infrared links are not really LANs, but they do provide a wireless serial link between computers and peripherals by replacing individual cables with light beams (see fig. 2.13). This technique makes it easy to transfer files between computers, such as between palmtop and desktop workstations. Several companies sell the links that interface to the serial port on your computer. Adaptec's AIRport 1000 and 2000 Infrared serial-port adapters, for example, eliminate the hassles of using cables or a set of floppy disks to transfer data from one machine to another. These products will transfer data between systems separated by up to three feet. This product line comes with a Windows-based file transfer with drag and drop functions. By the end of 1996, most portable computers sold should include this type of infrared link built into the unit.

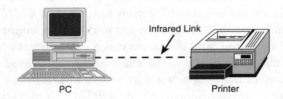

Figure 2.13
The "point and beam" technique for interfacing a computer to a printer.

Point-to-Point Infrared LAN System

Currently only one vendor, InfraLAN Technology, Inc., produces a product, InfraLAN, that implements a true point-to-point LAN system. InfraLAN consists of a pair of transducers, one for transmitting and one for receiving, that you configure as shown in figure 2.14. InfraLAN replaces the cable in the token-ring network with infrared light that can reach distances of up to 75 feet.

At each station, the InfraLAN interfaces with an IEEE 802.5 (token ring) interface board. Token-ring protocols ensure that only one station transmits at a time through the use of a token. The token, which is a distinctive group of bits, circulates the ring. If a station wishes to transmit data, the station must first capture the token, and then transmit its data. The capturing of the token ensures that no other station will transmit. The data circulates the ring and the appropriate destination will sense its address and process the data. Once finished, the sending station will forward the token to the next station downline.

Point-to-Point Infrared LAN System

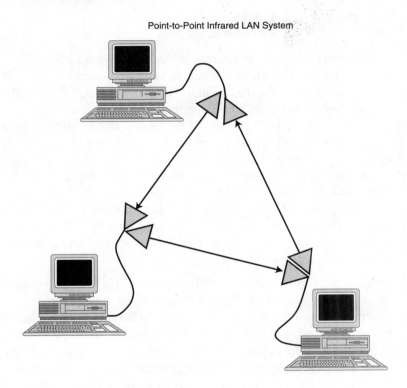

Figure 2.14
A point-to-point infrared LAN system.

The advantages of using InfraLAN is the performance and security it offers. Because of the focused infrared beam, the system can match performance requirements of either 4 or 16 Mbps token-ring protocols. InfraLAN is the only wireless LAN system on the market today that can support that type of performance. InfraLAN is also immune to electrical noise and is difficult to tap. Electrical signals do not interfere with the extremely high frequencies of infrared light, and an information thief would have to place himself within the path of the beam to receive the signal. The disadvantage with this approach, though, is that it does not accommodate mobility.

It might be suitable, however, in environments such as conference rooms, or in factories where electrical noise would interfere with radio signals.

Carrier Current LANs

A quasi-wireless LAN technique is the use of power lines as a medium for the transport of data. In the next year or so, you should begin seeing products that implement this approach. This technique is very similar to using an analog modem to communicate over telephone wires. Designers of the telephone system did not plan to accommodate computer communications, but people use modems everyday to communicate their data. The telephone system is capable of supporting analog signals with the range of 0 to 4 KHz. Telephone modems convert the computer's digital waveform to an analog signal within this range and transmit to the computer you choose. The modem at the distant end receives the "telephone signal" and converts the data back into a digital signal that is understood by the computer.

Power line circuits within your home and office provide enough bandwidth to support 1–2 Mbps data signals. Utility companies and others designed these circuits to carry 60 Hz alternating current typically at voltages of 110 volts. It is possible, then, to have a "power line" modem that interfaces a computer to the power circuitry (see fig. 2.15) The interface acts much like a telephone modem and converts the digital data within your computer to an analog signal for transmission through the electrical wires. The 110 volt alternating current in the circuit does not affect the signal (or vice versa) because the signals are at different frequencies. The interface has filters that will block the lower 60 Hz frequency from being received.

The advantages of this technique are ease of installation and low-cost products. A disadvantage of the power line approach is that the presence of electrical transformers, designed to electrically couple signals at 60 Hz, will block higher frequency data signals. Most homes and smaller facilities will not have this problem because usually only one side of the transformer is available; however, larger buildings, especially industrial centers, will have multiple electrical wire legs interconnected by transformers. The presence of transformers, therefore, will limit interconnectivity among sites.

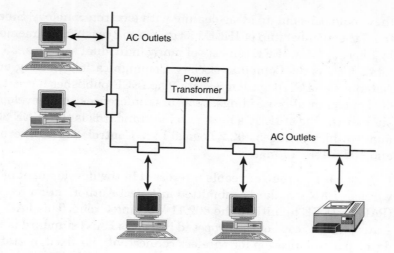

Figure 2.15
The carrier current LAN technique.

There are no products on the market today that implement the power line LAN approach, but Novell, Inc. is developing and promoting Powerline and targeting it toward home applications.

Wireless LAN Standards

As explained in Chapter 1, "Introduction to Wireless Networking," the lack of standards for wireless networks is causing some people to resist the implementation of wireless networks. Standard groups such as IEEE 802.11 and Infrared Data Association are developing standards that should ease people's minds.

IEEE 802.11

The lack of standards has been a significant issue with wireless networking. In response to this problem, the Institute for Electrical and Electronic Engineers (IEEE) has been involved in the development of wireless LAN standards for the last seven years. This effort is nearly complete—IEEE should finalize standards for wireless LANs in 1996 or 1997.

Who is IEEE? The IEEE is a non-profit professional organization founded by a group of engineers over a hundred years ago (1884) so they could consolidate ideas dealing with electrotechnology. Since then, the membership of IEEE has grown to over 320,000 members in 150 countries. IEEE consists of many individual societies, 35 in total, such as the Computer Society, Communications Society, and Antennas and Propagation Society. The IEEE publishes a great deal of technical proceedings, sponsors conferences, and develops standards. One of IEEE's most notable standards is the IEEE 802 family, which includes 802.2 Logical Link Control, 802.3 Ethernet, and 802.5 Token Ring.

In May 1991 a group of people interested in the development of wireless LAN standards submitted a Project Authorization Request (PAR) to IEEE to initiate the 802.11 Working Group. This PAR states, "...the scope of the proposed [wireless LAN] standard is to develop a specification for wireless connectivity for fixed, portable and moving stations within a local area." The PAR further says, "...the purpose of the standard is to provide wireless connectivity to automatic machinery and equipment or stations that require rapid deployment, which may be portable, handheld, or mounted on moving vehicles within a local area."

As with other 802 standards such as ethernet and token ring, the primary service of the 802.11 standard is to deliver MSDUs (MAC Service Data Units) between LLC (Logical Link Control) connections to the network. In other words, the 802.11 standard will define a method of transferring data frames between network adapters without wires.

In addition, the 802.11 standard will include:

○ Support of asynchronous and time-bounded delivery service

○ Continuity of service within extended areas

○ Accommodation of transmission rates between 1 and 20 Mbps

○ Support of most market applications

○ Multicast (including broadcast) services

○ Network management services

○ Registration and authentication services

According to the PAR, the 802.11 standard will provide Medium Access Control (MAC) and Physical Layer (PHY) specifications for

1 Mbps wireless connectivity between fixed, portable, and moving stations within a local area. In addition, a single MAC specification will support multiple PHYs using radio signals or infrared light for the transmission of data.

IEEE 802.11 meetings are open to anyone. The only requirement to attend is to pay dues which offset meeting expenses. Most of the active participants are representatives from companies developing wireless LAN components. The IEEE bylaws explain, though, that in order to vote on standards activities, you must participate in at least two out of four consecutive plenary meetings. Then, you must continue to attend meetings to maintain voting status. The 802.11 standards Working Group meets three times a year during the plenary sessions of the IEEE 802 and three times a year between plenary sessions.

The IEEE 802.11 consists of about 200 members, and membership falls into the following categories:

○ Voting Members, who have maintained voting status.

○ Nearly Members, who have participated in two sessions of meetings, one of which being a plenary session. Nearly members become voting members in the first session they attend following their qualification for nearly membership.

○ Aspirant Members, who have participated in one plenary or interim session meeting.

○ Sleeping Voting Members, those who were once voting members but have chosen to discontinue.

Companies, such as Proxim, Xircom, Windata, and many others, have been extremely active in forming standards within the 802.11 group. The final version of the 802.11 standard will make it an official standard, enabling the cross-vendor interoperability of wireless LAN components. As a result, most companies developing wireless LAN products today will certainly migrate their products to comply with 802.11.

Two subgroups comprise the 802.11 Working Group—the PHY Sub-Group and the MAC Sub-Group. The PHY Sub-Group of 802.11 is developing the physical layer of the standard. The PHY group has the following objectives: define the physical layer, develop a channel model, and develop conformance tests. The Working Group decided in July of 1992 to concentrate its radio frequency studies on the 2.4 GHz spread spectrum ISM band. This band is available license-free in most parts of the world. The 802.11 PHY Sub-Group established two ad-hoc groups, one for direct sequence and one for frequency hopping spread spectrum.

The PHY Sub-Group's efforts will result in a standard specifying three physical media: 2.4 GHz direct sequence spread spectrum, 2.4 GHz frequency-hopping spread spectrum, and diffused infrared. This will allow you to purchase 802.11 wireless LAN adapters from different vendors for a particular media type and be assured they will interoperate. Products that use one of these media types, such as direct sequence, however, will not interoperate with products using a different media, such as frequency hopping. In addition, the standard will not address roaming because it falls outside the scope of the architectural layers that the 802.11 standard addresses. You cannot expect one vendor's wireless access point to work with a different vendor's access point. This means you will need to standardize on one particular type of access point.

The MAC Sub-Group is concentrating on the MAC (Medium Access Control) layer of a network's architecture. The MAC layer defines the protocol that allows multiple stations to share the bandwidth of a common medium. The 802.11 group is developing a MAC layer that supports both asynchronous and Time Bounded Services (TBS).

To ensure interoperability with existing standards, the 802.11 Working Group is developing a standard that will be compatible with other existing 802 standards, such as:

○ IEEE 802 Functional Requirements

○ IEEE 802.2 MAC Service Definition

○ IEEE 802.1-A Overview and Architecture

○ IEEE 802.1-B LAN/MAN Management

○ IEEE 802.1-D Transparent Bridges

○ IEEE 802.1-F Guidelines for the Development of Layer Management Standards

○ IEEE 802.10 Secure Data Exchange

> **NOTE**
>
> For information on IEEE 802.11 proceedings, you can order specific documents through the IEEE 802 Document Order Service at 800-678-4333.

Infrared Light LAN Standards

The Infrared Data Association (IrDA) is a group of more than 80 computer and telecommunications hardware and software firms including Hewlett-Packard, AMP, Apple Computer, AST, Compaq, Dell, IBM, Intel, Lexmark, Motorola, National Semiconductor, Northern Telecom, Novell, Photonics, and Sharp. IrDA has adopted a standard covering three levels of a network's architecture: Serial Infrared Physical Layer Link (IrDASIR), Ir Link Access Protocol (IrLAP), and Ir Link Management and Transport Protocols (IrLMP). This standard specifies a 115.2 Kbps point-to-point infrared transmission between computers, laptops, printers, and fax machines. Other higher speed standards of 1.15 Mbps and 4 Mbps that will be more suitable for backups and offline storage are currently being considered by IrDA.

Migrating Existing Wireless LAN Products to 802.11

A common question that people ask is, "Will my existing wireless LAN hardware be able to interface with the upcoming 802.11 standard?" Most wireless LAN vendors, especially those that have been active in the development of 802.11, are planning to migrate their products to the new standard. These companies have been laying the seeds for 802.11 compatibility. Proxim, for example, has been heavily involved with the development of 802.11, shaping many aspects of the standard while developing RANGELAN2. Many of the ideas implemented in RANGELAN2 have been included in the proposed 802.11 standard and vice versa. RANGELAN2 and 802.11, for example, share a common contention-based media access scheme called Carrier Sense Multiple Access with Collision Avoidance (CSMA/CA). Also, 802.11 utilizes a dual data rate approach sponsored by Proxim, called 4FSK/2FSK, that maximizes both range and throughput. This mechanism allows networks to run at high speeds at short and medium range and "fall back" to a lower speed at long range. The point is that some vendors, such as Proxim, will have an easy time modifying their product to match 802.11 specifications. Owners of existing wireless LAN products, however, will probably have to interface these devices to 802.11 components through access points, as shown in figure 2.16.

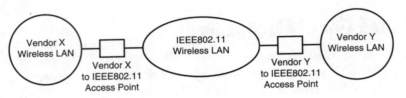

Figure 2.16
The interconnection of existing wireless LANs and an IEEE 802.11 compliant LAN.

Wireless LAN Case Studies

Many companies implement wireless LANs to support needs for wireless connectivity to computing resources. The following sections cover some of these examples.

TGI Friday's, Inc.[2]

Restaurant employees must juggle a great deal of information, such as lists of waiting customers, sizes of each party, seating preferences, table availability, orders, activity at each table, service requests, and bill payments. In addition, employees in this environment are always on the go, seating patrons, taking orders, and delivering drinks and food. If you multiply this information by the 40 to 80 tables in an average restaurant, plus the random elements of patron walk-ins and party size, you will have a good idea of the problems restaurant employees face in tracking information.

Some restaurant chains are trying to accommodate these employees with computer technology. Rock Systems, a Dallas-based hardware and software firm specializing in restaurant systems, assisted TGI Friday's, Inc. with implementing a system that reduces customer wait times. Rock Systems produces a table management software package called ProHost, which coordinates information from a variety of locations in the restaurant. This system works best if the employees carry a handheld computer that can communicate via a wireless network.

Rock Systems recently implemented such a system using ProHost and Proxim's RangeLAN2/PCMCIA at a TGI Friday's in Tallahassee, Florida. When guests arrive at the door, they give their name to the person working the wait list on a Fujitsu handheld computer. Their name is recorded and transmitted to the main ProHost terminal, which keeps track of the dining room and lets the host know what tables are available and the capacity of each. When a table of the appropriate size becomes available, the patron's name is automatically highlighted and the hostess walks the party to their seats. This function, as well as others, significantly reduces customer wait times.

American Red Cross[3]

When disaster strikes, the American Red Cross Disaster Service operates like a huge mobile warehouse—on a moment's notice, setting up locations for receiving and storing thousands of pallets of food, supplies, and equipment, and efficiently distributing those supplies to disaster victims. Often, these operations take place under extreme conditions, such as heavy storms, power and telephone outages, and floods. When relief needs have been met, they must be shut down quickly and the equipment made ready for immediate deployment to a new disaster site.

Most Americans are unaware of the fundamental logistical problems these activities present. Space suitable for warehousing must be quickly obtained from the often small pool of structurally safe buildings available for donation or leasing. After the Los Angeles Northridge earthquake of 1993, for example, Sears Roebuck donated the use of a one million square foot warehouse that served as the central Red Cross facility for the entire L.A. basin. By contrast, the relief operation deployed in response to the midwestern floods of 1993 and 1994 required numerous warehouses in several states due to the large geographical area affected.

The key requirements identified for this new automated system were mobility, reliability, ease-of-use by staff and volunteer workers, and the capability to provide six to eight hours of continuous battery operation (with battery swaps) in the event of power failure. The system needed to be capable of tracking everything from perishables and water to equipment such as fax machines, cellular phones, and tables and chairs. The system had to be capable of locally maintaining warehouse data and also transmitting that data to a central logistics database at the local disaster operational headquarters. The Red Cross central logistics database enables the volunteers to provide a current inventory of all relief material on hand for the entire operation.

Simple setup and ease-of-use were prime factors in the system design, both to increase overall warehouse efficiency and in response to the nature of the workforce at the warehouses. From a communications standpoint, the system had to be wireless with a range capable of communicating across large areas, through walls, and over high stacks of relief supplies. Dauphin's pen-capable DTR-2s were selected for the handheld portion of the system, which consists of 2–4 IBM Thinkpad 755s (one as a database server), a Canon portable printer, a Radix portable barcode printer, and 1–4 handheld DTR-2s outfitted with Maxtor PCMCIA hard disk drives for data backup and initial system load. The key to tying this all together was Proxim's RangeLAN2/PCMCIA wireless LAN cards.

Once the system was in place, all items entering the warehouse were barcoded on the fly at the pallet/lot level. These barcodes greatly increased the efficiency of receiving, shipping, and physical inventory operations. Nonexpendables and serialized items were barcoded individually to allow continuous tracking of their assignments and eventual return to the warehouse. Spectra-Physics laser barcode scanners attached to the keyboard ports of the Thinkpad and DTR-2 handhelds were used to scan Red Cross UPC codes on incoming food items.

The advantage of using a 486-based handheld PC as a data collection unit was its capability to run the same application software on all components of the system. The Dauphin units and the IBM Thinkpad notebooks operated as standard workstations on Proxim's wireless LAN when in radio range. In addition, any notebook or handheld PC could be quickly configured to be the server should a critical hardware failure occur.

Austin Regional Clinic[4]

In this age of outpatient services and cost-conscious providers, efficient, well-run medical clinics are quickly becoming the leading alternative to hospital-based treatment. Indeed, the large number of patients seen everyday at a medium-sized medical clinic are enough to strain the resources of many medical professionals and their staffs. To ease their jobs and expedite insurance billing so costs can be quickly recovered, these medical centers are increasingly turning to innovative forms of technology for answers. Austin Regional Clinic, a collection of 19 multi-specialty clinics employing 130 medical professionals who treat hundreds of patients a day in the Austin, Texas area, is a prime example.

Until recently, the volumes of paper-based notes on treatments and services rendered to these patients presented a serious billing problem. Specifically, the doctor's notes had to be transcribed into a billing system, checked, and corrected—all at significant cost— before being passed on to insurance companies as bills for service. This system created additional expense and delayed billing at least two weeks from the date of treatment.

To reduce this delay and minimize the paperwork and costs associated with it, the clinic decided two years ago to automate this process by giving medical professionals mobile, handheld computers to record patient services as they happened. Before the plan could be implemented, however, a logistical problem arose: how to download the patient service information to the clinic's central server. One possible method involved extending the clinic's existing Novell Netware 3.12 wire-based local area network. The other method would extend the existing wired infrastructure by adding wireless LAN adapters to existing computers that could transmit data to the central server instantaneously.

Austin Regional Clinic chose the wireless route because it saves the cost of pulling cable and the maintenance issues associated with this. At the same time, the wireless system enables medical professionals to walk from room to room while still providing full connectivity to the server.

Austin's answer was a Grid Pad portable, pen-based computer fitted with Proxim, Inc.'s RangeLAN2/PCMCIA wireless LAN adapter. RangeLAN2/PCMCIA provides data rates of 1.6 Mbps over 15 independent channels, for a total network bandwidth of 24 Mbps in any physical space.

Armed with these wireless mobile systems, medical professionals can now record office visits, lab work, X-rays, and other treatments directly to a digital format. The RangeLAN2/PCMCIA card downloads this information directly to a RangeLAN2/ISA card in the clinic's server. The data is then transferred by a Novell NetWare 3.12 LAN to the corporate database. Billing is executed by the corporate server within 24 hours of the patient encounter.

Methodist Hospital[5]

Indianapolis's Methodist Hospital installed Proxim's RangeLAN2 wireless communications technology to enable faster patient intake in the emergency room. The new system gives medical staff the ability to take patients straight to the treatment rooms, giving them immediate treatment and more privacy in divulging insurance information and medical problems.

Methodist Hospital is an 1100-bed private hospital with a 45-bed emergency room that features a state-approved, Level-1 trauma center. Approximately 85,000 to 90,000 patients pass through the hospital's emergency room each year, and many are in need of immediate treatment. For these patients, there is no time to wait while a registration clerk collects information about the reason for the visit, type of insurance coverage, and personal health history. Sometimes this information must be recorded as the patient is being transported to a room, or even while they are being treated.

To expedite the registration process, Methodist Hospital worked with Datacom for Business, a Champaign, Ill.-based value added reseller. As a result, the hospital remodeled its 65,000-square-foot emergency room, eliminating all but two registration tables and replacing the rest with Compaq Contura notebook computers equipped with wireless LAN adapters. Now, patients can go directly to treatment rooms where registration clerks gather the necessary intake data and enter it into the database on the host computer.

Methodist Hospital's wireless communications are made possible by RangeLAN2/PCMCIA adapters from Proxim, Inc. The wireless PCMCIA adapters allow the Compaq notebook computers running TN3270 terminal emulation to access the clinic's existing wired client/server network or communicate on a peer-to-peer basis with other mobile systems within the same clinic site. The adapters operate at an average power output of about 100 milliwatts and use advanced power management to minimize the drain on the mobile systems' batteries. RangeLAN2 provides easy access to standard wired LAN environments, including the hospital's existing TCP/IP network. This is accomplished through the use of three Proxim RangeLAN2/Access Points, which act as wireless bridges and allow mobile users anywhere in the emergency room to send information to the TELNET LAN server. The terminal emulation is then transferred in real-time over a TCP/IP enterprise backbone to the hospital's database in the mainframe computer.

Disaster Recovery/Manufacturing

Autoliv, a Swedish company founded in 1953, manufactures and distributes auto safety equipment such as airbags, seat belts, and child seats. With factories in virtually all major European manufacturing countries, Autoliv has positioned itself as one of the largest distributors of safety equipment in Europe. In July 1995, a fire in the Vargarda facility caused extensive damage to over 6,000 square meters (65,000 square feet) of production and storage facilities. Furthermore, existing network cabling and a majority of the manufacturing terminals were damaged in the blaze. Given the severity of the fire and the company's dedication to customer satisfaction, a solution was needed to rapidly fulfill existing customer orders.

By incorporating a configuration of WaveLAN-equipped PCs and WavePOINT access points, Autoliv was able to successfully and rapidly rebound from the disaster. Within a matter of days the WaveLAN wireless local area network was installed. The WaveLAN solution helps track approximately 800,000 pieces of equipment delivered to customers daily.

In addition to the swift and easy installation, WaveLAN is providing Autoliv the flexibility and mobility Autoliv lacked in a traditional wired LAN environment. Inventory control, materials tracking, and delivery information are available online from various locations on the production floor. WaveLAN's 2 Mbps data speed helps to further increase productivity by handling this critical information easier than the 9400 baud speed of the old system. WaveLAN's cost effectiveness is also becoming apparent by helping Autoliv save on reconfiguration and rewiring costs that average nearly 3500 Swedish kronas ($500) per terminal.

INDUSTRY: Airline Industry

In the competitive world of airline travel, computer reservation systems such as Apollo, SAABRE, PARS, and WorldSpan help airlines and travel agencies with the planning, scheduling, and ticketing of flights throughout the world. When one of the largest airlines in the United States sought a solution to extend the reach of its computer reservation system, it sought to offer travel agencies a cost-effective, turn-key computer solution.

WaveLAN was chosen by the airline for plug-and-play installation of its computer reservation system terminals. Travel agency LANs are preconfigured in a centralized location then distributed to travel agency locations nationwide. Travel agencies can easily install the computer system without waiting for wiring to be installed.

This plug-and-play approach has resulted in significant cost savings for the airline and its travel agencies. The airline estimates that it has saved nearly 30 percent on initial installation costs and nearly 35 percent on the total cost of ownership versus the traditional wired alternative. Other benefits include configuration flexibility, ease of relocation, and quick installation—no local expertise is needed.

Younkers, Inc.

Younkers, Inc., the Des Moines, Iowa-based chain of fashion department stores, has always emphasized the need to focus attention on individual stores. Clear and constant communication between headquarters and the stores' sales associates is essential, since Younkers bases its buying decisions on detailed customer profiles, which differ geographically. When the company acquired H.C.

Prange department stores in 1993, the new branches' means of communication—their point-of-sale (POS) systems—were incompatible with Younkers' existing equipment. Younkers decided to absorb the cost of standardizing the POS systems at all 53 stores, which meant rewiring for twisted-pair cable. Their original estimate was $400 per terminal.

Taking into consideration the need for high-speed transmission and quick installation (not to mention on-site obstacles to wiring), Younkers decided to pursue a wireless network alternative. The company incorporated wireless WaveLAN cards onto its NCR 7052 POS terminals, linked through a WavePOINT wire-to-wireless bridge, to an in-store processor that can perform such specialized database functions as customer gift registration. The in-store networks are in turn connected to the headquarters' mainframe, which handles credit authorizations, merchandise locations, and other applications. The decision to adopt a WaveLAN-based system allowed Younkers to meet (and even exceed) its anticipated goals of greater processing power, timely roll-out, low cost, and flexibility. The company has assembled a system that enables employees to instantly access host-based information from virtually any store or remote site using versatile terminals that can be moved anywhere at any time. Thanks to the mobility of WaveLAN workstations, Younkers has even been able to reduce the number of POS terminals throughout its stores. The retailer also avoided the expense and disruption of installing twisted-pair wire in 53 separate buildings.

National Gallery of Art

The National Gallery of Art in Washington, D.C., boasts one of the world's largest and most diverse collections of fine art. The Gallery's shop plays an important role in raising funds to maintain the collection—each year, visitors purchase more than 200,000 exhibition catalogues, art books, and more than 1.5 million printed reproductions. In addition to the main shop, temporary retail setups are established throughout the museum's two buildings for special exhibitions, which generally remain in place for a few weeks or months at a time. However, the buildings' marble floors (which cover two city blocks) and other physical obstructions make setting up these temporary locations impractical, since registers must be hardwired into the Gallery's ethernet network for credit card authorization, price look-up, and end-of-day balancing.

The Gallery has simplified the creation of these special retail stations by installing WaveLAN on its AT&T 7450 terminals. These terminals can then be wheeled to the exhibition site and connected wirelessly to the museum's ethernet network, which consists of several LANs responsible for everything from accounting to building automation. No wires are run between terminals, which provides management with more flexibility in terms of where the retail operation for a particular exhibition is placed. When a cashier needs to make a price inquiry or credit-card authorization, the request is made wirelessly from the issuing terminal to the nearest WavePOINT access device, several of which are installed throughout the buildings. WavePOINT communicates to the server in the Gallery's office, which supplies the pricing information, or dials out for authorization, and then responds to the POS terminal.

WaveLAN has eliminated the problem of exposed wiring, which has considerably reduced terminal down time as well as the hazard of scattered wires. Moreover, set-up time for retail stations is much shorter, since the WavePOINT access devices are pre-installed and furnish broad coverage. All employees have to do is roll their terminals to the desired location and set them for the appropriate WavePOINT address. WaveLAN makes wireless transaction processing pervasive throughout the museum, while at the same time providing terminals with continuous access to the most recently updated data on all of the museum's various databases.

Sydney Observatory

Sydney Observatory, part of Australia's Museum of Applied Arts and Sciences, has provided star-gazing to astronomy enthusiasts for nearly 140 years. Built in 1858, the Observatory has since been classified by the National Trust as one of Australia's historical buildings. When the Observatory began investigating ways to share these views of space with a much broader audience, the obvious solution was to download images to multiple PCs and large screens via a local area network. Due to the historical nature of the building, however, cabling was not an option. Very thick sandstone walls and historic plaster ceilings could not be easily drilled into, and strings of cable would have been unsightly and unsafe to the public. The only solution was a wireless LAN—WaveLAN.

WaveLAN is installed in each of the Observatory's eight PCs and the network server. Telescopic images are downloaded from the Internet or from electronic cameras housed in the Observatory Dome's telescopes. These images are then displayed on the various PCs for individual viewing or on larger monitors for group viewing.

In addition to improving access to images and network information, the Observatory has the flexibility to change the locations of the PCs within the building. WaveLAN has made office work and administrative tasks much simpler and more efficient. Mr. George Rossi, the network administrator, points out that "AT&T's wireless solution has effectively solved the Observatory's networking problems."

---

[1] Reprinted from the WaveLAN Design Guide by permission from Lucent Technologies

[2] Reprinted by permission from Proxim, Inc.

[3] Reprinted by permission from Proxim, Inc.

[4] Reprinted by permission from Proxim, Inc.

[5] Reprinted by permission from Proxim, Inc.

Wireless Metropolitan Area Networks (MANs)

Are you thinking about connecting network sites within the same metropolitan area? Organizations often have requirements for communications between facilities in a semi-local area, such as a city block or metropolitan area. A hospital, for example, might consist of several buildings within the same general area, separated by streets and rivers. A utility company also might have multiple service centers and office buildings within a metropolitan area.

Traditionally, companies utilize physical media—such as buried metallic wire or optical fiber, or leased 56 Kbps or T1 circuits—to provide necessary connections. These forms of media, however, might require a great deal of installation time and can result in expensive monthly service fees. A cable installation between sites several thousand feet apart can cost thousands of dollars or more, and leasing fees can easily be hundreds of dollars per month. In some cases, leased communications lines might not even be available.

This chapter explains the applications of wireless MANs and covers the following wireless MAN types and applicable products:

○ Radio-Based Wireless MANs

○ Laser-Based Wireless MANs

Wireless MANs use technologies very similar to wireless LANs described in Chapter 2, "Wireless Local Area Networks (LANs);" therefore, this chapter will concentrate on technological aspects differing from wireless LANs.

Before getting into the technologies and products, you should understand what drives the need for wireless MAN connectivity. A wireless MAN can provide communications links between buildings in these situations, avoiding the costly installation of cabling or leasing fees and the downtime associated with system failures. The city of Macon, Georgia, for example, uses Cylink's wireless products to provide links for traffic control at ten consecutive intersections over a four-mile stretch of state highway. This system avoids the installation of wiring along the roadway. Other organizations, such as hospitals and government centers, use wireless MAN components to avoid digging trenches and routing cabling around rivers and roads.

A wireless MAN can result in tangible cost savings rather quickly. ATCO Products in Ferris, Texas, for example, installed a wireless MAN between its existing plant and a new plant under construction about 12 miles away. ATCO will recover all the installation costs in less than two years and will then begin a positive return on its investment. Many other companies are also realizing these types of benefits.

Another application of wireless MAN components is to facilitate a backup in case a primary leased line becomes inoperative. An organization can store the wireless equipment at a strategic location, such as the computing center, to have on hand if a primary link goes down. If a primary link fails, an organization can quickly deploy a wireless link to restore operations.

A wireless MAN, as illustrated in figure 3.1, utilizes either radio waves or infrared light as a transport for the transmission of data up to 30 miles. These systems work in a point-to-point configuration, much like that of leased lines. Wireless MANs interface easily and match the data rates of existing LANs. The cost of connecting two sites in a wireless MAN ranges from $1,500 to $20,000, depending on data rate and type of transport. The following sections explain each of these techniques.

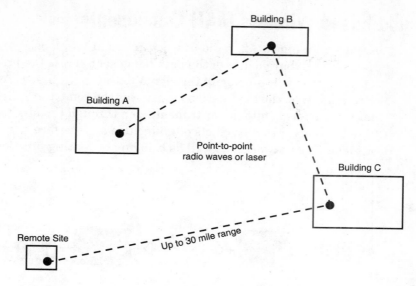

Figure 3.1
The wireless MAN concept.

Radio-based Wireless MANs

A radio-based wireless MAN is currently the most common method for providing connectivity within a metropolitan area. These products have highly directional antennas to focus the signal power in a narrow beam, maximizing the transmission distance. As a result, spread spectrum products operating under one watt of power can reach single-hop transmission distances of 30 miles. The actual transmission distance of a particular product, though, is dependent on environmental conditions. Rain, for example, causes resistance to the propagation of radio signals, decreasing the effective range.

Radio-based wireless MAN data rates are four to five Mbps for the shorter range products operating over two to three mile links. Most products operating over a 30 mile link, however, will transmit at much lower data rates. Wireless MAN products use either spread spectrum or narrow band modulation.

Radio-based Wireless MAN Components

As shown in figure 3.2, radio-based wireless MANs consist of transceivers that modulate the data being sent across the link with a carrier that will propagate the signal to the opposite site. As with wireless LANs, the modulation transposes the computer's digital data into a form suitable for transmission through the air. Radio-based wireless MAN products often include an interface to ethernet or token ring networks, as well as bridging or routing functionality.

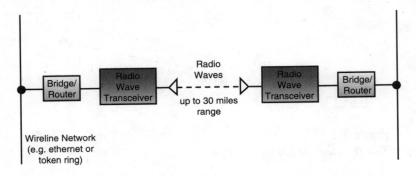

Figure 3.2
A radio-based wireless MAN.

Wireless MAN bridges, also called wireless remote network bridges, segment data traffic by filtering each packet according to its final destination address. This form of segmentation blocks packets from crossing the wireless link unless they need to reach a destination on the opposite side. As with local bridges, this makes better use of bandwidth and increases network performance. The router versions of these products work very much like traditional routers that are dependent on physical media—they forward packets based on the final destination address. This makes it possible to create a more intelligent network having alternate routes. In fact, a collection of these components would constitute the functionality of a WAN limited to a large metropolitan area. The greatest difference, though, is that the wireless MAN will not support mobile users—it only provides wireless connections between fixed sites.

Spread Spectrum Wireless MANs

As with wireless LANs, wireless MANs using spread spectrum in the ISM bands do not require user licensing with the FCC (refer to Chapter 2, "Wireless Local Area Networks," for an explanation of spread spectrum). The advantage of not dealing with licenses is easy and rapid installation. For instance, the installation time of spread spectrum products can take a few hours, saving the two month wait for FCC licensing.

Spread spectrum resists interference from traditional narrow band radio systems, enabling systems using both modulation techniques to exist in the same space. The only signals that are likely to cause serious interference originate from other spread spectrum devices. The disadvantage, then, is the possible interference with others operating similar wireless MANs nearby. With wireless LANs, interference is normally not a problem because the radio waves are kept indoors, within the confines and management of the organization. Radio waves traveling between buildings, perhaps across a large city, will be beyond the organization's jurisdiction and control, possibly receiving interference from other unknown systems. The owning organizations and users of the wireless MAN equipment, therefore, will probably not be able to do anything about the interference.

Many companies sell spread spectrum wireless MANs. The following are examples of spread spectrum-based wireless MAN products:

Proxim's RangeLINK

Proxim's RangeLINK family of high speed wireless bridges connects LANs in buildings separated by up to three miles. RangeLINK operates in the 2.4 GHz band at 1.6Mbps using frequency-hopping spread-spectrum technology. The RangeLINK Series 1500 is for multipoint solutions and enables you to have a central bridge (RangeLINK 1510) that connects multiple remote facilities within a range of one mile from the central site. Each remote facility must use a RangeLINK 1520 or 1530. The RangeLINK Series 2000 is for high-speed, longer-distance (up to three miles) solutions.

Persoft's Intersect Remote Bridge

Persoft's Intersect Remote Bridge uses spread spectrum radio frequency operating with the 902 MHz and 2.4 GHz ISM bands to filter packets between points at two locations. The Intersect Remote Bridge has a data transmission rate of 2.0 Mbps with a maximum range of five miles. This product can interface with either ethernet or token ring LANs.

Solectek's AIRLAN

Solectek has a product, the AIRLAN Bridge Plus, which operates in the 902 MHz ISM frequency band and delivers data at 2 Mbps. AIRLAN provides connectivity between ethernets located up to three miles apart.

Solectek's AIRLAN/Router 200 routes Internet Protocol (IP) and Internetwork Packet Exchange (IPX) packets within MAN environments. AIRLAN/Router 200 combines multiprotocol routing with wireless bridging to provide an internetworking solution at 2 Mbps for LANs up to 25 miles apart. The AIRLAN/Router 200 software, co-developed by Cisco Systems, increases network availability and performance by providing broadcast control, protocol mediation, and network security using optional DES encryption. The AIRLAN/Router 200 utilizes spread spectrum radio in the 2.4 GHz ISM band. The AIRLAN/Router 200T is compatible with IEEE 802.5 Token Ring, and the AIRLAN/Router 200E is compatible with IEEE 802.3 Ethernet. Antennas mount on the roof of the building and require a line-of-sight path to the destination.

Cylink's AirLink Bridge

Cylink's AirLink Bridge functions as a remote ethernet bridge that filters packets at full, wired ethernet speed and forwards only the appropriate packets across wireless links up to 20 miles long. In addition, the link is full-duplex, so there are no end-to-end collisions that degrade throughput and cause re-transmission. The AirLink bridge uses direct sequence spread spectrum radios tuned to 2.4 GHz.

LANNAIR's EthAirBridge

LANNAIR produces a wireless bridge called EthAirBridge, which interfaces Ethernets over a range of 15 miles. It operates in the 2.4

GHz ISM band and delivers data at 76.8 Kbps. An optional 4:1 data compression board increases performance to 307.2 Kbps. In addition, an RS-232/V.24 or V.35 interface is available that can support a dial-up back-up link. The RS-232 interface supports data rates from 4.8 to 19.2 Kbps, and the V.35 interface supports data rates up to 2.048 Mbps.

Windata's AirPort

Windata, Inc. makes a system called AirPort, which operates at 5.7 Mbps data throughput. AirPort reaches this relatively high speed by separating the send and receive channels onto different radio bands and using error correction. The AirPort I system has a range of 1,000 feet, and AirPort II can send signals within a 0.9 mile radius.

SpreadNet's Wireless Link

SpreadNet's product, Wireless Link, provides throughput up to 2 Mbps. The Wireless Link product is available in configurations that support transmission distances of short range (5 miles), medium range (6–25 miles), and long range (70 miles using repeaters).

Narrow Band Wireless MAN

The FCC and comparable regulatory agencies in other countries regulate the use of narrow band frequencies. This regulation offers both an advantage and disadvantage. The FCC licenses each user site to operate on an assigned frequency, often a 25 KHz slice of bandwidth, which gives the user specific rights for operating on the assigned frequency at a specific location. If interference occurs, for example, the FCC will intervene and issue an order for the interfering source to cease operations. This is especially advantageous when operating wireless MANs in areas having a great deal of operating radio-based systems.

The disadvantage is that the licensing process can take two or three months to complete. You must complete an application, usually with the help of a frequency consultant, and submit it to the FCC for approval. Thus, you can't be in a hurry to establish the wireless links. Plus, you will probably have to coordinate with the FCC when making changes to the wireless MAN topology.

Multipoint Networks sells one of the very few wireless MAN products that operates on narrow band frequencies. Multipoint produces both point-to-point and point-to-multipoint narrow band wireless MAN systems. All Multipoint products operate within the 400-512 MHz and 820-960 MHz frequencies and have a range of 30 miles or more. The point-to-point product, called Radio Area Network (RAN), operates in full-duplex mode ranging in data rates from 9.6 to 125 Kbps. The main applications of RAN are for the replacement of leased lines and wireline modems within metropolitan areas.

Multipoint's transparent point-to-multipoint product, waveNET 2500, is an intelligent hub that supports X.25 connectivity to remote locations. In a point-of-sale or Automatic Teller Machine (ATM) application, you can use Multipoint's LaunchPAD to interface the POS terminal to the waveNET 1000. WaveNET's Radio Access Protocol (RAP) manages the data traffic between LaunchPADs and the hub.

Laser-based Wireless MANs

Another class of wireless MANs utilizes laser light as a carrier for data transmission. A laser, which is now a common term for Light Amplification by Stimulated Emission of Radiation, contains a substance in which applied electricity causes the majority of its atoms or molecules to be in an excited energy state. As a result, the laser emits coherent light at a precise wavelength in a narrow beam. Most laser MANs utilize lasers that produce infrared light. (Refer to Chapter 2, "Wireless Local Area Networks," for concepts of infrared light.)

As with other wireless techniques, a laser modem in this type of system modulates the data with a light signal to produce a light beam capable of transmitting data. With light, these data rates can be extremely high. Most laser links can easily handle Ethernet (10 Mbps), 4/16 Mbps token ring, and higher data rates. Figure 3.3 illustrates a laser MAN.

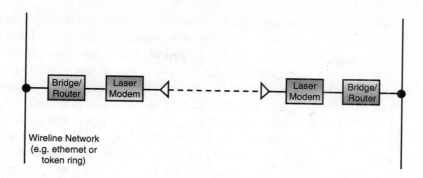

Figure 3.3
A laser-based wireless MAN.

To maintain safe operation, laser links typically range less than a mile. These devices comply with the Center for Devices and Radiological Health standards and most operate at class III, which can cause eye damage under some circumstances. Distances much longer than a mile are possible, but you would have to increase the power to a level that would damage buildings and injure living things.

Weather is also an influence on the transmission distance of laser systems. A nice, clear day with very little smog will support the one mile operating distance. Snow, rain, fog, smog, and dust, however, cause attenuation, which could limit the effective range to a half mile or less. A fairly heavy rain shower (3-4 inches per hour), for example, will introduce approximately 6 dB of attenuation per kilometer. As a result, you need to plan the link according to potential changes in weather.

Why use laser-based MAN technology over radio types? One reason is the need for high speed data transmission. A laser MAN system is the only way to effectively sustain 10Mbps and higher data rates, which may be necessary for supporting the transfer of CAD (Computer Assisted Drawing) files and X-ray images. Also, you do not have to obtain FCC licensing. The FCC doesn't manage frequencies above 300 Ghz; therefore, you can set up a laser system as quickly as you can set up a license-free spread spectrum radio system.

When using a laser, very few other systems can cause interference. Even at high microwave frequencies, radio signals are far from the spectral location of laser light, eliminating the possibility of interference from these systems. Also, an interfering laser beam is unlikely because it would have to be pointed directly at your receiving site. Sure, someone might do this to purposely jam your system, but otherwise it won't occur. Sunlight consists of approximately 60 percent infrared light and can cause interference. The rising or setting sun might emit rays of light at an angle that the laser transducers can receive, causing interference in the early morning and late afternoon. Therefore, an organization should avoid placing laser links with an east-west orientation. Generally, laser-based MANs are highly resistant to interference. Thus, laser links might be the best solution in a city full of radio-based devices, especially for applications where you must minimize downtime.

To accommodate a line-of-sight path between source and destination, the best place to install the laser link is on top of a building or tower. This avoids objects blocking the beam, which can cause a disruption of operation. Birds are generally not a problem because they can see infrared light and will usually avoid the beam. A bird flying through the beam, however, will cause a momentary interruption. If this occurs, higher level protocols such as ethernet or token ring will trigger a retransmission of the data. The infrared beam will not harm the bird.

Laser-based systems offer more privacy than radio links. Someone wanting to receive the laser data signal would have to physically place himself directly in the beam's path (see fig. 3.4). Also, the eavesdropper would have to capture the light to obtain the data, significantly attenuating or completely disrupting the signal at the true destination. This means he would have to put himself next to the laser modem at either end of the link by standing on top of the building or climbing to the top of a tower. Physical security, such as fences and guards, can effectively eliminate this type of sabotage.

A small number of companies sells laser-based wireless MAN components. Here's an overview of the leading products:

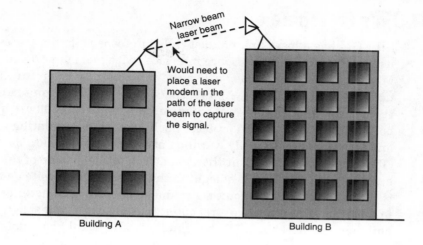

Figure 3.4
The difficulty in capturing data from a laser-based wireless MAN.

LCI's LACE

LCI (Laser Communications Incorporated) has a product line called Laser Atmospheric Communication Equipment (LACE). LACE is a laser-based system that operates in the near infrared region of the spectrum at a wavelength of 820 nanometers. LCI recommends a maximum range of 1 Km (3340 feet) for LACE, but the actual transmission distance depends on atmospheric conditions, such as rain and smog, and desired data rates. Faster data rates limit the range in order to keep errors at an acceptable level (better than 10^{-9} error rate). LCI's wireless MAN products interface with a variety of protocols, such as 10 Mbps Ethernet, 4 or 16 Mbps, 2.0 Mbps V.35, T-1/E-1, RS 232/422/423, 34 Mbps E-3, 45 Mbps T-3, 51 Mbps SONET or ATM, 100 Mbps Fast Ethernet 802.3U, 100 Mbps FDDI, and 155 Mbps SONET or ATM. The units that house the laser modems are weather-tight and mount on the top of a building or tripod. LACE will work through glass; however, the glass surface will reduce the light intensity by approximately four percent. In order for the light to penetrate the glass, the beam must be nearly perpendicular to the pane of glass, and the glass cannot have any infrared reflecting coating.

SILCOM's Freespace

SILCOM has a laser link product called Freespace that operates using infrared light. Freespace interfaces with and supports 10 Mbps Ethernet, full-duplex Ethernet, or 4/16 Mbps Token Ring at up to 300 meters (1000 feet). Freespace has a built-in, uninterruptable power supply to keep the link running up to three hours if power fails. The unit also protects itself against accumulating ice and snow by automatically warming at temperatures below 34°F (1°C). Freespace has a unit that you attach to the outside of the building with a clear line of sight to the unit at the opposite end of the link. A built-in telescopic sight and signal strength meter eases installation. You can place a repeater, router, or bridge inside the building and attach it to the head unit via an optical fiber cable.

Wireless MAN Case Studies

Many companies are beginning to reap the benefits of wireless networks by seeing better returns on initial investments of wireless MAN hardware in comparison to other forms of connectivity, such as optical fiber installations and 56 Kbps and T1 leased services. The following case studies show how some companies are using wireless MAN products.

Wireless ATM/POS Data Communications Network[1]

First Security Bank of New Mexico (formerly First National Bank in Albuquerque) has one of the largest wireless ATM (Automated Teller Machine) networks in the western US with over eighty digital wireless modems in operation. The bank has four point-to-multipoint wireless systems currently installed throughout the metropolitan Albuquerque area. Two of the systems each serve approximately 20 ATMs at branch offices. The third system serves about 20 stand-alone ATMs at convenience stores, supermarkets, hotels, sport facilities, travel centers, and so on. The fourth system connects an electronic benefit transfer (EBT) system between the bank's data center and eight point-of-sale (POS) terminals in supermarkets throughout the city.

One of the systems accommodates a recreational vehicle (RV) equipped with two ATMs. Believed to be the first of its kind in the world, the RV is available for special events such as conventions, the annual Hot Air Balloon Fiesta, and state and county fairs where no other banking facility is available.

Figure 3.5 illustrates the system configuration. This system uses the RAN products available from Multipoint Networks.

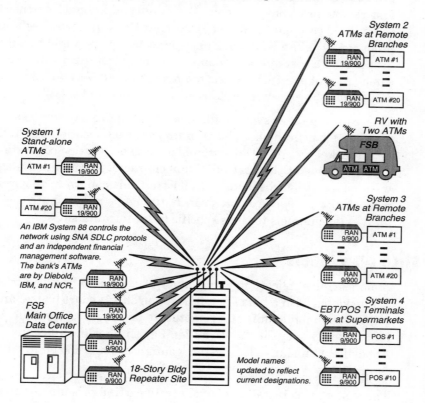

Figure 3.5
A wireless ATM/POS data communications network.

Wireless Branch Banking Network [2]

Franklin National Bank is located in rural northeast Texas. The bank implemented wireless technology to provide a high-performance data transmission facility for teller terminals between the host data center at the main office in Mt. Vernon and new remote branches established in outlying population centers. The branches are 20 to 30 miles from Mt. Vernon. The system initially connected four remote branches. The bank is planning additional branches for the future. Through the use of Multipoint Network's wireless MAN products, Franklin National Bank connected the new additional branches in a cost-effective manner and provided them with the same efficient service as at the main bank in Mt. Vernon. Leased line costs would have been prohibitive in this same application.

The resulting teller system, illustrated in Figure 3.6, services multiple asynchronous teller terminals at each location. An X.25 PAD (packet assembler/disassembler) at each location converts the asynchronous bit stream (individual characters) from the multiple terminals into synchronous data for transmission over the wireless modem link. At each remote branch, the PADs redistribute the individual channels to their specific teller terminals.

Justice on Wheels[3]

A pioneering application, which involves a pair of Cylink's AirLink 128 modems with a video monitor at one site (a court house) and a roll-around mobile unit at a remote site (a jail), is making life easier and saving money for the Vermilion Parish Sheriff's office in Abbeville, Louisiana. The system configuration is shown in figure 3.7.

The system was installed to avoid transporting prisoners to the Court House for hearings and arraignments. There are many benefits of being able to communicate with the prisoners face to face, but considerable pressure exists to make a single visit suffice because someone must escort the prisoner nine miles from the jail to the Court House, hampering the security of the public and the prisoner. Also, scheduling conflicts often arise, forcing the escort to take the prisoner back to the jail to wait for the rescheduled appointment.

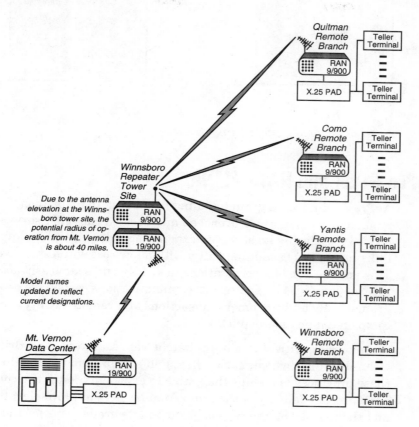

Figure 3.6
A wireless branch banking network.

Installation of the system only took one day by placing an AirLink 128 in both the Vermilion Parish Court House and the jail and then installing directional antennas on the roofs of both buildings. When people in the courthouse want to interview a particular inmate, jail personnel wheel a video conferencing cart—equipped with a camera, TV microphone/video codec, and AirLink 128—to the prisoner. This saves the Parish court system a great deal of time and money.

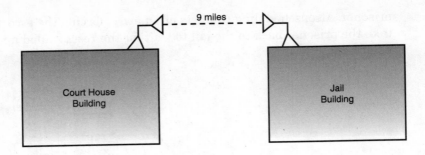

Figure 3.7
A wireless video conferencing system.

AgPro Grain and RANGELINK[4]

AgPro Grain, Inc., a grain handling firm based in Saskatoon, Saskatchewan, trimmed overhead and boosted customer service levels by replacing its slow and expensive modem-based dial-up links to a nearby satellite facility with a low-cost wireless network. This enables remote workstations to swiftly and seamlessly access the company's LAN. The result: communications costs have dropped dramatically and customer transactions at a remote facility are completed much more quickly.

AgPro Grain, with offices in Saskatchewan, Manitoba, and British Columbia, buys wheat, barley, flax, and canola from farmers in these provinces and ships the grains to export markets. The company accepts delivery of the grain from farmers at grain elevators and then, based on the type of grain and its grade, pays for the delivery and logs the transaction into a settlement system.

In Saskatoon, farmers deliver large quantities of grain to an inland terminal that also houses business offices; smaller deliveries are made to a smaller grain elevator, called Saskatoon A, 700 feet away. With 12 staff members at the main business office, and only two at Saskatoon A, implementing separate networks at each site to support the settlement system was not cost-effective.

Faced with the problems of optimizing transaction speed for farmers at Saskatoon A, enhancing overall network operation control, minimizing capital expenditures, and controlling ongoing communication costs, AgPro Grain looked for a way to integrate workers at this remote facility with a new LAN implemented at the main business office. The most obvious solution was a leased telephone line. Because traffic is so light between the two sites, however, it could not be cost-justified. Similarly, linking the sites with fiber optic or coaxial cable was also too expensive, due in part to the costs of obtaining permits to lay cable under the railroad lines that separate the facilities. The company then considered utilizing the same frame relay network used to link offices and elevators at its Manitoba facilities, but the costs of this option would have been $600 per month and low data speeds.

With no conventional communications alternatives meeting its needs, AgPro Grain began a market search for a low-cost, easy-to-install wireless link between the two sites that would not be affected by extreme climatic conditions—including temperatures that vary from 40 degrees Celsius in the summer to minus 40 in the winter. AgPro decided to use Proxim's RangeLINK product. RangeLINK has been operating flawlessly throughout Saskatoon's harsh winter climate at speeds that enable transactions at Saskatoon A to be completed four times faster than with the modem-based dial-up system that AgPro Grain previously used.

---

[1] Reprinted by permission from Multipoint Networks
[2] Reprinted by permission from Multipoint Networks
[3] Reprinted by permission from Cylink
[4] Reprinted by permission from Proxim

Wireless Wide Area Networks (WANs)

Do your professionals on the road need to have access to e-mail and other computing resources at their home office? The traditional solution is to equip the person's portable computer with a wireline modem and access on-line services and other resources via the Plain Old Telephone System (POTS). The user can interface his modem to the telephone line and dial into the services and resources the user wishes to utilize. This solution works well, assuming the professional has access to a telephone line. Most hotels and office facilities can accommodate a temporary POTS connection; however, other places do not. For instance, many travelers spend a great deal of time in airports waiting for plane connections that are often delayed. Unfortunately, there is no place to plug your computer into the POTS at an airport. In addition, you don't typically find POTS connections at archeological dig sites or environmental survey sites. Some hotels and office buildings also might not have a telephone line that you can use. For these situations, a wireless WAN might be the solution for effectively connecting people to the computing resources they need.

It's important to understand the various wireless WAN technologies and services before deciding on a solution. This chapter describes the following technologies, services, and products that provide wireless WAN connectivity:

○ Packet Radio WANs

○ Analog Cellular WANs

○ Cellular Digital Packet Data WANs

○ WAN-Related Paging Services

○ Satellite Communications

○ Meteor Burst Communications

○ Combining Location Devices with Wireless WANs

○ Wireless WAN Case Studies

Packet Radio WANs

A packet radio WAN uses packet switching to move data from one
location to another. In general, a user wishing to utilize packet
radio networking purchases a radio modem for his portable computer and leases access to a packet-based wireless network from a
service provider such as ARDIS or RAM Mobile Data. The main
advantage of packet radio is its ability to economically and efficiently transfer short bursts of data that you might find in systems
such as short messaging, dispatch, data entry, and remote monitoring. However, packet radio systems do not yet have worldwide
coverage; today, coverage is limited to large cities. In the next few
years, this coverage should be near 100%.

Packet Radio Architecture

A packet radio network performs functions relating to the physical,
data link, and network layers of the OSI reference model. Therefore, this type of network performs routing and provides a physical
medium, synchronization, and error control on links residing
between nodes or routers. If more than one hop is necessary to
transfer data packets from source to destination, then the intermediate packet radio nodes relay the data packets closer to the destination, very much like a router does in a traditional wire-based
WAN.

Packet Radio Components

To utilize a packet radio network, the user must equip his notebook
or palmtop computer with a radio modem, applicable communications and application software, and lease a packet radio network
service from one of several service providers (see fig. 4.1).

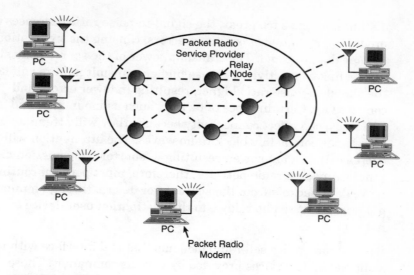

Figure 4.1
Packet radio network components.

Packet Radio Modems

A radio modem provides an interface between end-user devices and radio relays, using air as the medium. These modems are capable of transmitting and receiving radio waves at user throughput rates up to 20 Kbps. They usually do not require licensing, making it easy to move them from one location to another. Most radio modems transmit omnidirectional radiation patterns.

As long as connectivity exists, a pair of radio modems establishes a channel for data transmission between sites. The main condition for proper connectivity is that the destination must be able to correctly receive data from the source at a specified minimum data rate. For example, if the reception of data on a particular channel results in a number of bit-errors exceeding the maximum error rate for that link, connectivity is lost.

Due to node separation, transmit power, and irregular terrain, most packet radio networks are not able to maintain full connectivity. That is, not every user access device and radio relay node have connectivity with each other. Node separation affects the connectivity of a radio network because the power of a radio signal decreases exponentially as the distance between the nodes increases. If the distance becomes too great, the signal-to-noise ratio decreases and produces too many transmission errors, causing the two stations to become disconnected. The transmit power of the source node also affects link connectivity because higher transmit powers will keep the signal-to-noise ratio higher, resulting in fewer errors and connectivity. Certain types of terrain, such as mountains and buildings, can affect connectivity because they will attenuate and sometimes completely block radio waves. The attenuation will decrease the signal power, resulting in shorter transmission distances. A packet radio network, therefore, must perform routing to move data packets from the source user device, through a number of intermediate radio relays, to the destination user device or network.

Several companies sell radio modems that can interface with packet radio network services provided by various companies. These modems are service provider-specific and use different frequencies. A later section in this chapter describes several packet radio network service providers, as well as the modems that work with those services.

Relay Nodes

The radio relay nodes, which implement a routing protocol that maintains the optimum routes for the routing tables, forward packets closer to the destination. The routing table contains an entry for each possible destination relay node (see fig. 4.2).

Destination Address	Next Hop (Relay Node)
Node 1	Node 1
Node 2	Node 2
Node 3	Node 3
Node 4	Node 4
Node 5	Node 5
Node 6	Node 6

Figure 4.2
A relay node routing table.

Packet Radio Operation

To carry packets from source to destination, a packet radio network must do the following:

1. Transmit data packets
2. Update routing tables at the relay nodes

Transmitting Data Packets

When the application software at a user's appliance requests the transfer of data through the network, communications software prepares the data for transmission by wrapping it with a header that primarily contains the destination address and some trailer bits that represent a checksum. The relay nodes use the address to determine whether to forward the packet to the next relay node or send it to the final destination. A receiving node utilizes the checksum to detect if the packet encountered any transmission errors. If errors are not present, the receiver sends back an acknowledgment; otherwise, the source retransmits the packet.

Each station, whether it is the user's access device or a relay node, uses a carrier sense protocol to access the shared air and radio medium. Ethernet and radio-based wireless LANs operate in a very similar pattern. The primary difference is that a packet radio network operates in a partially connected instead of fully connected topology (see fig. 4.3). The propagation boundary of a particular node defines that node's operating range. Nodes A and C, for example, are within each other's propagation boundary; therefore, they can communicate directly with each other.

A packet radio station wishing to send a data packet must first listen to determine if another station is transmitting. If no other transmission is heard, then the sending station will transmit the packet in a broadcast mode using its omnidirectional antenna. With most packet radio networks, the first station to receive the packet will be the neighboring relay node. This relay will look in its routing table to determine which node to send the packet to next, based on the final destination address. If the destination is located within range, the relay node will broadcast the packet again and the destination will receive it. If the final destination is not close by, the relay node will obtain the address and broadcast the packet to the next relay node closer to the destination. This process will continue until the packet reaches the destination.

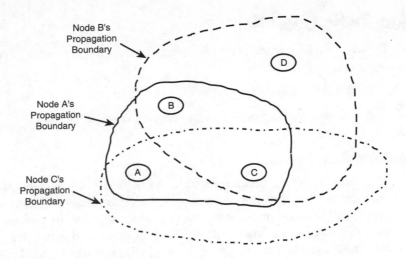

Figure 4.3
The topology of a packet radio network.

Updating Routing Tables

As with wireline WANs, a packet radio topology may change over time. Relay nodes might become inoperative, new relay nodes might appear, and atmospheric conditions, such as rain and sunspots, might affect radio connectivity between stations. These connectivity changes can alter the topology; therefore, the efficient operation of a packet radio network depends on an effective routing protocol capable of updating the routing tables at each relay node.

There are several approaches to updating routing tables; however, the distributed protocol is most common with the Internet and packet radio networks. Wireline WANs utilize routing protocols such as Routing Information Protocol (RIP) and Open Shortest Path First (OSPF). These protocols enable each router within the network to gain a complete picture of the network's topology. Packet radio routing protocols have the same goal.

Specifically, distributed routing protocols make it possible for each relay node to determine the next path to send a packet by operating as follows: each relay node periodically sends a status packet that announces its presence to all neighboring relay nodes within range. Each node, then, periodically learns the presence of its immediate neighbors. A router can use this information to update its routing table. When a relay node sends its status message, it also sends a copy of its routing table. Each relay node also sets timers for each neighbor, and if the relay node does not receive a status message within a certain time period, the relay node will delete the neighbor from the routing table. Other relay nodes will hear of the deletion via the periodic status messages. Through this process, each relay node will eventually obtain a complete picture of the network in terms of connectivity.

Packet Radio Service Providers

Several packet radio service providers have constructed networks that implement the radio relay nodes. The process of establishing packet radio networking is to lease the service from one of several packet radio access providers. This service is very economical, costing pennies to send small e-mail messages. These companies usually supply the software for no charge; however, users must purchase a radio modem.

The following sections provide an overview of several packet radio network providers.

ARDIS

ARDIS is a company that leases access to their wireless WAN, which is based on packet radio technology. The ARDIS network covers 410 top metropolitan areas in the United States, Puerto Rico, and the U.S. Virgin Islands. This network encompasses more than 80 percent of the population and 90 percent of the business areas. ARDIS uses two different protocols—MDC4800 at a data speed of 4,800 bps, and Radio Data-Link Access Protocol (RDLAP) at a speed of 19,200 bps. ARDIS was originally developed for IBM service technicians who worked indoors. As a result, ARDIS was designed to have good in-building coverage.

In addition to basic wireless WAN interconnectivity, ARDIS offers the following wireless WAN applications that enable you to communicate with other ARDIS and Internet users, as well as implement some specific applications:

○ **ARDIS PersonalMessaging**. Enables users to send and receive messages to ARDIS and Internet users.

○ **ServiceExpress**. A wireless communications solution that makes it easier for field service organizations to incorporate wireless communications. ServiceExpress provides field service engineers with access to corporate information systems using handheld computers. ARDIS offers this service with a low monthly payment per field engineer. The payment covers the cost of all hardware, ARDIS air time, training, equipment repair, maintenance, and a 24-hour help desk. In most cases, ServiceExpress will work with a company's existing field service management software.

○ **TransportationExpress**. Combines all the components of a wireless dispatch solution including dispatch hardware and software, ARDIS wireless air time, handheld wireless computers for pickup and delivery drivers, project management, installation, training, and a 24-hour help desk.

IBM Wireless Modem for ARDIS

To interface with ARDIS, you can utilize the IBM Wireless Modem for ARDIS, which has a PCMCIA Type II form factor. This card can be installed internally in an IBM ThinkPad or externally with the optional External Battery Assist and tethered radio transceiver. The product also includes an external transceiver that you connect to the card via a cable. IBM also offers the modem in a form that fits into the floppy drive of a ThinkPad. The modem operates with a power output of 0.8 watts.

RAM Mobile Data

RAM Mobile Data is another alternative for wireless WAN connectivity. RAM covers the top 266 United States metropolitan areas plus airports and major transportation corridors. This includes more than 92 percent of the urban business population. RAM is committed to providing 100 percent coverage through RAM's Strategic Network initiative. For example, a RAM user will be able to maintain connectivity through the use of circuit-switched cellular and satellite when traveling beyond RAM's wireless WAN coverage area. In order to provide the Strategic Network, RAM has linked up with satellite service providers and telecommunications carriers to ensure a complete and seamless solution.

RAM Mobile Data utilizes Mobitex technology, which is an established and proven packet radio system for the transmission of data. Mobitex infrastructure was originally developed by L.M. Ericsson in 1983 in cooperation with Sweden's Postal Telegraph and Telephone national communications authority. Mobitex today is seen as a de facto standard for packet-based wireless WANs. The Mobitex Operators Association (MOA) now manages Mobitex standards. There are currently 17 other wireless WANs throughout the world that use Mobitex technology, with more planned for 14 other countries.

RAM Mobile owns and operates its network and has licensed frequencies from the FCC with the 896–901 MHz and 935–940 MHz bands. RAM parcels these frequencies to give each metropolitan area up to 30 channels. Each channel supports a data transmission speed of 8 Kbps. RAM provides wireless connectivity to mobile users within specific service areas, then uses a regional switch to tie the service areas together.

The RAM Mobile Data network provides nationwide, transparent and seamless roaming. Subscribers do not have to manually intervene when moving from one service area to another. Each service area uses a different frequency, and the RAM modem automatically locates the best available channel and local switch. In addition, RAM does not charge fees for roaming.

RAM offers an e-mail service for sending and receiving messages with other RAM and Internet users. The service also enables you to utilize many other e-mail products. In fact, over 80% of all e-mail products have been RAM-enabled, meaning they will interface with the RAM wireless network. For example, Lotus cc:Mail, Microsoft Mail Remote for Windows, and Novell Groupwise support the RAM network.

IBM Wireless Modem for RAM

To interface with RAM Mobile's network, you will need to obtain a compatible modem. IBM offers its Wireless Modem for RAM, which is housed in a PCMCIA Type III card. The modem operates from a self-contained battery and uses the portable computer's battery for a continuous trickle-charge. The modem operates at 2 watts of output power and minimizes power consumption by using sleep and battery saving modes during operation. The IBM RAM modem was manufactured by Ericsson. The RAM card comes with ZAP-it software, which provides e-mail access to the Internet and fax capabilities.

Metricom

Another radio-based wireless WAN is Metricom's Ricochet, which provides two-way multi-user data communications. For a relatively low price, Ricochet offers user throughput rates of 28.8 Mbps with unlimited data transmissions. The Ricochet network enables personal computer users to wirelessly access the Internet, America Online, Compuserve, and most other dial-up services from anywhere within a covered area.

> **N O T E**
>
> Although Ricochet's current coverage is limited to the southern San Francisco area, Metricom is making large deals to expand their network. In 1995 Metricom agreed to form a joint venture with PepData, Inc., a subsidiary of Potomac Electric Power Company (PEPCO) of Washington, D.C., to deploy, own, and operate a wireless network providing data communications services to four million potential customers throughout the metropolitan D.C. area. As part of the joint venture agreement, Metricom is installing a network of poletop radios on street lights, power poles, and buildings to provide Metricom's Ricochet wireless data communications service to residential, business, and governmental users. Metricom will supply the Ricochet technology, install the components, and operate the network.

The architecture of Ricochet is shown in figure 4.4. The concept is to use wireless access points and network radios (radio relays) on top of light poles and buildings approximately one half mile apart to facilitate connectivity between user appliances. The radios operate in the license-free (902-928 MHz) portion of the radio spectrum using frequency hopping spread spectrum modulation. Ricochet interfaces with the RS-232 port of a user appliance via Ricochet's wireless radio modem, allowing it to work with any application that can interface with standard wireline modems. Ricochet implements standard AT modem command set as well as SLIP (Serial Line Internet Protocol) and PPP (Point-to-Point Protocol) protocols for direct serial access to the Internet or any private TCP/IP-based LAN. The modem transmits the user's data packets to the nearest poletop radio. The data packets then move through the mesh from poletop to poletop until they reach their final destination.

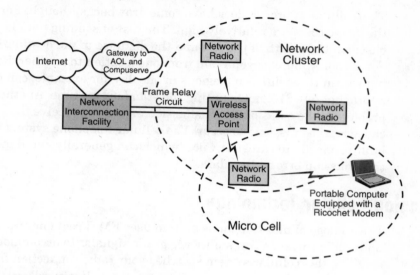

Figure 4.4
The architecture of Metricom's Ricochet.

Analog Cellular WANs

The analog cellular telephone system has made it possible for millions of people to make phone calls using portable "cell" phones while away from the home or office. Many who start using cell phones can't imagine life without them. The general idea of an analog cellular WAN is to make use of the cell phone's mobility and employ it as a means of transferring data, much like the use of traditional wire-based Plain Old Telephone Service (POTS). You can use the analog cellular phone system to dial-in to your corporate network to access applications and send e-mail, just like you can when using POTS.

If you need to send large files, such as engineering drawings, this cellular phone approach might be the way to go. It provides data rates up to 28.8 Kbps, which is similar to conventional POTS telephone modems. The idea of this technology is to connect your computer to a cellular telephone via a modem and then with a remote system through a dialup connection. This provides a relatively easy way to obtain wireless data transfer wherever cellular telephone service exists, which covers most of the world.

This cellular approach does have some drawbacks, though. For one, the usage costs are relatively high. The cost of sending data is based on the length of the call and the distance to the destination. You also pay more as you roam from one location to another. Roaming within three different regions in one day, for example, can cost approximately $10, not counting the standard air time. Another problem is the occasional transmission errors you receive that will cause retransmissions to occur. The cellular telephone system was built primarily to transmit voice, which can generally stand more transmission errors than data.

Analog Cellular Technology

The analog cellular telephone system uses FM (Frequency Modulation) radio waves to transmit voice grade signals. To accommodate mobility, this cellular system switches your radio connection from one cell to another as you move between areas. Every cell within the network has a transmission tower that links mobile callers to a Mobile Telephone Switching Office (MTSO). The MTSO, which is owned and operated by the cellular carrier in your area, provides a connection to the public switched telephone network. Each cell covers several miles. Figure 4.5 illustrates the general topology of the cellular telephone system.

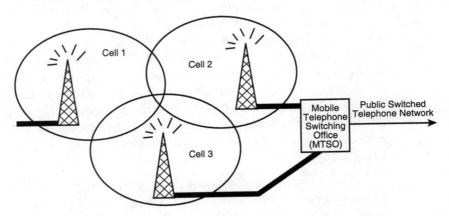

Figure 4.5
The general topology of circuit switched cellular telephone system.

Most modems that operate over wireline telephone services will also interface and interoperate with cellular phones; however, modem software optimized to work with cellular phones minimize battery usage. MobileWare, for example, is software that enables you to communicate over cellular or regular wireline phones. MobileWare minimizes the cost of cellular service usage by making a connection only during actual information transmissions. MobileWare prepares the information for transmission and then transmits whenever possible. Thus, you can prepare correspondence such as e-mail messages and faxes while on an airplane, then MobileWare will send the information after you land. MobileWare also "mobilizes" the Lotus cc:Mail, meaning it enables you to effectively utilize cc:Mail on a wireless connection.

NOTE

A major problem is that no standard connector exists to interface your portable computer's serial interface to a cellular phone. Therefore, be sure you can find a cable to connect between the modem and the phone.

Cellular Digital Packet Data (CDPD) WANs

To establish a dedicated wireless data network for mobile users, a consortium of companies in the United States developed the Cellular Digital Packet Data (CDPD) standard. CDPD overlays the conventional analog cellular telephone system, using a channel-hopping technique to transmit data in short bursts during idle times in cellular channels. CDPD operates full duplex, meaning simultaneous transmission in both directions, in the 800 and 900 MHz frequency bands, offering data rates up to 19.2 Kbps.

What's the advantage of using CDPD versus analog cellular systems? Recall that the main advantage of the analog cellular system is widespread coverage. Since CDPD piggybacks on this system, it will also provide nearly worldwide coverage. The main plus with CDPD, though, is that it uses digital signals, making it possible to enhance the transmission of data. With digital signaling, it's possible to encrypt the data stream and provide easier error control.

CDPD is a robust protocol that is connectionless and utilizes Reed-Salomon forward error correction (FEC). FEC is an error control technique that corrects errors at the receiver without asking the source to retransmit the errored packet. Security with CDPD, which is accomplished using an encrypted key-passing technique, is very good. Also with CDPD, you only pay for the amount of data that is actually sent, which is less than the amount you would spend on an analog cellular call if sending the same data.

CDPD is available today only in large cities within the United States, but it is spreading to other areas as well. However, many visualize CDPD as an interim solution until digital cellular telephone service becomes available.

CDPD Design Goals and Objectives

The CDPD Forum is an industry association with over 90 companies interested in developing and promoting CDPD products and services. The Forum's main mission is to develop a technical standard for CDPD, as well as develop the market place and promote the technology.

As part of the development of the CDPD System Specification, the companies agreed upon these design objectives:

○ Maintain compatibility with existing data networks, technology, and applications

○ Support a wide range of present and future data network services and facilities

○ Provide maximal use of existing data network technologies and minimize the impact on existing end-user appliances

○ Allow a phased deployment strategy in terms of basic connectivity, security and accounting, network management, and application services

○ Provide services that relate to mobile and portable situations

○ Support equipment from multiple vendors

○ Provide seamless service to all subscribers

○ Protect the subscriber's identity

○ Protect the subscriber's data from eavesdropping

○ Protect the CDPD network against fraudulent use

○ Support conservative use of the airlink interface

○ Support use of a wide variety of mobile situations

The CDPD Airlink Interface specification defines all procedures and protocols necessary to allow effective use of existing analog cellular channels for data communications. The initial CDPD System Specification was published in July 1993. Release 1.1 provided some updates and was published in January 1995.

CDPD Architecture

Figure 4.6 illustrates the architecture of the CDPD system. The Mobile Data Base Station (MDBS) defines a radio cell that interfaces the Mobile End System (M-ES), such as a portable computer with a CDPD modem, with the Mobile Data Intermediate System (MD-IS). The MD-ISs provide mobility management services for the CDPD network. The MDBS acts as a bridge between the wireless protocols of the M-ES and the landline protocols of the MD-IS. Therefore, the MDBS decodes the data received from mobile devices, reconstructs the data frames, and transfers them to the MD-IS. At most cellular telephone base sites, digital communications lines tie the MD-ISs back to the cellular telephone system's MTSO. The CDPD architecture includes mobility management and internetworking between separate CDPD network providers, resulting in seamless operations as you move between different cells and providers.

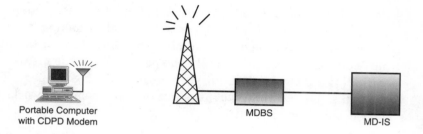

Figure 4.6
The architecture of the CDPD system.

An M-ES wishing to send data uses the Digital Sense Multiple Access (DSMA) protocol to share radio channels with other M-ESs. This protocol is similar to the common ethernet protocol used in LANs. DSMA, though, has both a forward and reverse channel that supports full-duplex operation. MDBSs transmit on the forward band, and M-ESs transmit on the reverse band.

The forward channel includes transmission of Busy/Idle and Decode Status channel indicators. If an MDBS detects a transmission on the reverse band (coming from a mobile device), it transmits a signal with the Busy/Idle channel status indicator set. An M-ES wishing to send data will first check for this transmission and indicator before transmitting. The M-ES will not transmit until the indicator is set to Idle. The MDBS sends a Decode Status indicator to indicate whether it successfully decoded the received data transmission. If the M-ES does not receive a positive acknowledgment, it will attempt to retransmit the data.

To utilize CDPD, you will need to lease the service through your local provider and purchase a CDPD modem. An example of a CDPD modem is IBM's Cellular Modem, which is housed in two packages—a PCMCIA Type II card and an external transceiver that you connect to the card via a cable. IBM also offers the modem in a tray-mounted version that is installed inside the IBM ThinkPad's floppy drive. The Cellular Modem supports three modes of operation: analog voice, analog data, and CDPD.

WAN-Related Paging Services

Paging has been around for years. You often see doctors, chaplains, and service personnel with pagers on their belts, making it possible for someone to call the beeper number and alert the user to call a particular number. Traditionally, the paged person then responds by finding a phone and calling the person initiating the page. Today, companies are beginning to introduce two-way pager networks that offer networking capabilities.

The use of paging systems, especially those that support two-way transmissions, is an effective method for providing WAN services. Pagers have been very successful because of their light weight, long battery life, and ability to work indoors. These are strong selling points, especially if you require highly mobile communications that must work practically anywhere.

Two-Way Pager Networks

Companies have been working hard to enhance their paging infrastructures to two-way data services. SkyTel, for example, recently launched the first two-way paging and messaging service. The SkyTel 2-Way was the first narrow band Personal Communications Service (PCS), which enabled people to respond to and automatically acknowledge pages.

SkyTel's two-way paging network, illustrated in figure 4.7, was designed for rapid messaging. Outbound messages from the Network Operations Center (NOC) use a different path to reach subscribers than messages use on the return path. Mtel (SkyTel's parent company) owns nationwide narrow band licenses consisting of three 50 KHz forward paths and five 12.5 KHz reverse channels. These channels are within the 901 and 940 MHz frequency band.

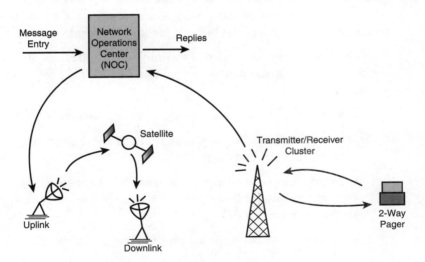

Figure 4.7
SkyTel's two-way paging network.

SkyTel's network uses Motorola's ReFLEX 50 narrow band PCS protocol that provides 25.6 Kbps throughput for numeric, alphanumeric, and binary data types. Messages are sent to the individual pagers over a 50 Khz channel using a satellite-based system and high-powered transmitters. Responses from the pager are sent at a lower rate (9,600 Kbps) to conserve the pager's battery power. A series of small receivers deployed around the high-power transmitters collect these responses and send them back to the originator via the telephone system.

The NOC is an intelligent control site and tracks all network activity. The NOC receives all incoming messages originating from sources such as touch-tone phones, operator assistance, computers, the Internet, and network providers. The NOC then distributes the messages, collects confirmations and responses, and transmits them back to the message origination. The NOC can also route responses to one-way paging devices.

SkyTel's 2-Way Paging and Wireless Messaging Services

SkyTel, in conjunction with their partners, has developed a variety of two-way messaging products based on their two-way paging system. SkyTel's 2-Way Pager is pocket-sized, weighs only 5.5 ounces, has 100K of memory, and features a 4-line LCD display. The pager runs on a single AAA battery for up to several weeks.

SkyTel's 2-Way paging service is available in 1,300 cities, including the top fifty U.S. metropolitan areas, and includes the following features:

- Receive text messages of up to 500 characters in length

- Respond immediately to messages with short-predefined, multiple-choice responses

- Compose and send responses of up to 95 characters using SkyTel's Palmtop Messenger (with a cable connection to Packard's palmtop PC)

- Initiate messages composed on the Hewlett Packard's palmtop PC through SkyTel's 2-Way pager to other SkyTel 2-Way pagers

- Receive numeric messages and notification of voice messages

- Receive news and information services, such as SkyNews and SkyQuote

- Guaranteed delivery of messages through store and forward messaging techniques

One-Way Paging Services

Today, there is a mixed bag of one-way paging services that provide wireless WAN functionality. CompuServe, for example, offers wireless pager capability, which enables subscribers to send e-mail messages from their PCs to other subscribers' pagers. CompuServe contracts with RPA, Inc., which acts as a hub and interconnects with every paging company in the United States. CompuServe's service enables subscribers to send e-mail messages directly to numeric and alphanumeric pagers. The maximum message length depends on the specific paging vendor, but most pagers support messages up to 80 characters long; some permit up to 240 characters. Another service enables users to receive notification via their pager when their CompuServe mailbox receives e-mail. Other online services such as Microsoft Network, Prodigy, and America Online are deploying similar services, too.

Socket Wireless Messaging Services (SWiMS)

Socket Wireless Messaging Services (SWiMS) is a suite of services administered by the National Dispatch Center and handled by Socket Communications. SWiMS receives messages or data sent to a personal 800 number over conventional phone lines and then routes messages to your pager. If someone wants to send you a message, he calls your personal 800 number using a phone, fax machine, or modem-equipped personal computer. For voice calls, a dispatch operator will answer and transcribe the message into text and send the message immediately to your pager. If someone sends you a voice mail or fax, a notice is sent to your pager and you can call into your account to listen to the voice mail or forward the fax to a particular fax machine. Also, someone can send e-mail to your pager's Internet address that includes your 800 number as part of the address (such as 8005551212@pagecard.com). SWiMS will convert the first 600 characters of the e-mail message into a page and transmit it to your pager.

SWiMS gives you the following coverage options:

○ **Local.** You can receive your messages in a single metropolitan area.

○ **Roaming.** Redirects your messages to one of more than 80 major metropolitan areas.

○ **Nationwide.** Covers over 200 major metropolitan markets across the United States. Some smaller cities are available only on nationwide coverage.

Satellite Communications

Most wireless WAN services, such as analog cellular, packet radio, CDPD, and paging networks, have fairly good coverage, but offer relatively low data rates. If you are looking for high speed transmission with complete worldwide coverage for your portable users, then satellite communications might be a good alternative.

The main issues with satellite systems are high costs and limited support for mobile users. The monthly service costs are not too bad, but initial equipment costs are very high. The portability aspect of satellite, mainly because of the antenna sizes and point-to-point uplink, requires users to setup the antenna dish and align it with the satellite before sending data.

Satellite System Components

Satellite systems today support transmission of video, voice, and data for a variety of companies that require global coverage. The components that make this possible are shown in figure 4.8. The satellite is a platform that hosts a series of transponders acting as signal repeaters. The transponders receive directed signals on the uplink from earth stations and broadcast the signals back to earth on a downlink frequency where users over a very wide area are able to receive the signal. A satellite in geostationary orbit has a 24 hour period at an altitude of 22,300 miles over the equator. The satellite takes one full day to orbit the Earth, which makes the satellite appear to look stationary from the Earth's surface. This enables the earth station antennas to remain fixed, not having to track the satellite. The geostationary orbit puts the satellite directly over the Earth's equator; therefore, earth stations in the Northern Hemisphere point their antennas toward the southern horizon, and stations near the equator, such as Bogotá, Columbia, point their antennas straight up into the sky. Some satellite systems, though, work at lower altitudes to lessen the amount of propagation delay. These lower altitude satellites are not geostationary and will

appear to move across the sky, requiring earth stations to have
tracking antennas.

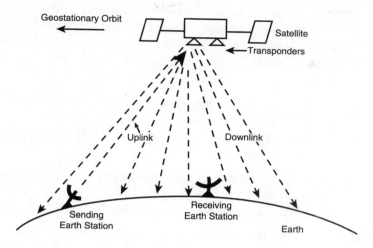

Figure 4.8
Satellite system components.

To utilize a satellite system, you need a satellite station that consists of an antenna, satellite transceiver, and an interface to your
computer. There are several companies that sell these components.
For example, California Microwave has a product called LYNXX
Transportable Inmarsat-B Earth Station that provides 64 Kbps
throughput for voice, video, fax, and data transmissions. LYNXX
can serve as a global wide area gateway into the Internet, as well as
corporate networks. The system has a 50 foot interface cable,
enabling you to mount the antenna outside and operate the computer from within a building.

Hughes Network Systems Inc.'s DirecPC

Another option for satellite networking is Hughes Network Systems
Inc.'s DirecPC, a PC-based satellite interface that enables users to
access the Internet, capture broadcasts, and write them directly to
the hard drive at speeds up to 2 Mbps. DirecPC services are based
on VSAT (Very Small Aperture Satellite Terminal) facilities, and an
access kit for PCs includes a 24-inch VSAT dish connected by cable
to an ISA-bus card. The kit also includes Windows-based software.

DirecPC has several subscriber services as follows:

○ **DirecPC Digital Package Delivery.** Distributes large files simultaneously to many locations. You can send the files on demand or pre-scheduled. To send data, the information provider transmits the information to the DirecPC Network Operation Center, which broadcasts the data via a 12 Mbps Ku-band satellite to one or more recipients.

○ **DirecPC Access Service.** A basic monthly subscription that provides the user access to a wide range of information services, such as online news, sports, financial information, and software ordering.

○ **DirecPC Multimedia.** A service enabling you to schedule the broadcast of desktop video, audio, or regularly transmitted information. This service uses DES encryption for security.

○ **DirecPC Turbo Internet.** A high speed, low cost Internet connection that uses DirecPC technology to receive data packets from the Internet.

The DirecPC 16-bit ISA adapter card enables a PC to receive high speed satellite data via the DirecPC's Network Operations Center. The adapter card provides DirecPC signal reception and DES encryption to prevent unauthorized access.

The 24-inch dish antenna can withstand severe weather, and you can mount it in a number of areas—sloped roof, vertical wall, ground pole, or exterior pipes. When installing DirecPC, allow about an hour to assemble and mount the antenna, install the ISA adapter card, and adjust the antenna to pick up the correct satellite signal.

ACT Network's SkyFrame

ACT Network's SkyFrame combines satellite modem and radio frequency (RF) terminal, multiplexer, and frame relay technology. Refer to Chapter 8, "Designing a Wireless Network," for a discussion on frame relay. SkyFrame makes frame relay services available to sites where it is not physically or economically feasible to use terrestrial-based frame relay. SkyFrame consists of a base unit that houses modulators, demodulators, voice/fax cards, LAN cards, and high-speed data cards. The modulator card includes a frame relay switch and a modulator that concentrates all packets received from the access ports and transmits the data. The demodulator receives the satellite signal and filters the received packets according to their addresses.

Comsat Mobile Communication's Planet 1

Many other companies are working on satellite solutions that support user needs for remote access to the Internet and corporate networks. Comsat Mobile Communications, for example, is working on a personal satellite communications system that will offer (in 1997) personal voice and data communications on a "seamless" global basis. Comsat calls this system Planet 1, and it will use a portable notebook-sized terminal. The system will incorporate functions such as global roaming, fax and voice-mail notification, paging, e-mail, and Internet access.

Planet 1 will provide secure communication channels via the Inmarsat-3 satellites. The transponders on the satellite will use spot beam technology that concentrates greater power onto the Earth's surface. This improves spectrum efficiency because the system provides frequency reuse and should minimize per-minute usage charges.

Meteor Burst Communications

Occasionally at night, you can see the burning trail of a meteor as it enters the Earth's atmosphere. Actually, billions of tiny microscopic meteors enter the atmosphere every day. As these meteors penetrate the ionosphere, they leave a trail of ionized gas. Meteor burst communications direct a radio wave, modulated with a data signal, at this ionized gas (see fig. 4.9). The radio signal reflects off the gas and is directed back to Earth in the form of a footprint that covers a large area of the Earth's surface, enabling long distance operation.

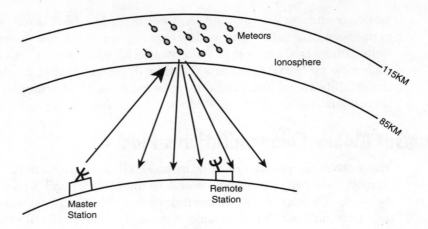

Figure 4.9
A meteor burst communications system.

A meteor burst communications system is advantageous because it can reach into remote areas where there is no packet radio or cellular network coverage. In addition, the implementation of a meteor burst system will generally cost less than leasing satellite service. These benefits make meteor burst systems well-suited for remote telemetry, water management, environmental monitoring, pipeline regulation, and oceanographic observation.

To use a meteor burst system, you must purchase and implement the equipment yourself. Meteor Communications Corporation (MCC) sells the MCC-520B Meteor Burst Master Station and the MCC-550C Remote Data Terminal that support data rates up to 8 Kbps over a range of 1,000 miles. These devices operate in the 40–50 MHz frequency range.

The Master Station is the main component in a meteor burst communications system. It controls the routing of messages and data from hundreds of Remote Data Terminals throughout the system. A Remote Data Terminal collects data from analog and digital sensor inputs and is designed for unattended and automatic operation. When the Remote Data Terminals receive a beam from a Master Station, they are able to send data containing information obtained from their sensors. Because of low power consumption, the Remote Data Terminals can operate unattended for a year or more.

Combining Location Devices with Wireless WANs

Location devices identify your position on the Earth's surface in terms of latitude and longitude. Traditionally, products that provide this type of information have been relatively expensive; therefore, they have mainly been used for providing navigation information to aircraft and ships. In the last few years, however, the size and cost of location devices have shrunk dramatically.

Location devices are based on Global Positioning System (GPS) technology—a worldwide, satellite-based radio navigation system providing three-dimensional position, velocity, and time information to users having GPS receivers anywhere on or near the surface of the Earth. Today, you can purchase a handheld position indicator for several hundred dollars. With this device and a good map, you're unlikely to get lost. GPS devices tell you your position on earth within a few meters. Plus, many products, such as mapping software, are starting to incorporate GPS technology.

The GPS was developed by the U.S. Department of Defense and provides two levels of service—a Standard Positioning Service (SPS) and a Precise Positioning Service (PPS). SPS is a positioning and timing service available to all GPS users as a continuous, worldwide service with no direct charge. SPS provides the capability to obtain horizontal positioning accuracy within 100 meters and vertical positioning accuracy within 140 meters. The PPS is a highly accurate service used by the military to obtain positioning, velocity, and timing information.

The GPS satellites operate on two L-band frequencies: L1 (1575.42 MHz) for SPS and L2 (1227.6 MHz) for PPS. The system uses spread spectrum modulation and provides a great deal of resistance to interference. Each satellite transmits a navigation message containing its orbital elements, clock behavior, system time, and status messages. A user's GPS receiver can determine its position by obtaining time information from three satellites. Altitude determinations require more than three satellites.

NOTE

The GPS consists of space, control, and user segments. The space segment consists of 24 satellites in six orbital planes. The satellites operate in circular 10,900 nautical mile (20,200 km) orbits at an inclination angle of 55 degrees and with a 12-hour period. The control segment consists of Monitor Stations and Ground Antennas. There are five Monitor Stations with one located in Hawaii, Kwajalein, Ascension Island, Diego Garcia, and Colorado Springs. There are three Ground Antennae, one at Ascension Island, Diego Garcia, and Kwajalein. The monitor stations gather range data by tracking all satellites within view. The user segment consists of antennas and receiver-processors, providing positioning, velocity, and precise timing to the user.

GPS/Wireless Applications

The combination of wireless WAN technologies and location devices offer some interesting applications. The joining of these technologies makes it possible for a mobile element to communicate its exact location to other elements. Many companies are combining CDPD and paging with the Global Positioning System (GPS) in their products, mainly for vehicle tracking.

ETE's MobileTrak

ETE has a product, MobileTrak, which is a portable, low-cost, Automatic Vehicle Locating (AVL) system that works with ETE's line of wireless communications products. MobileTrak combines wireless, two-way packet data communications with GPS to create a low cost, off-the-shelf, AVL system. The MobileTrak system depicts vehicle location on a full color map on either a laptop, desktop PC, or Macintosh that can be integrated with wireless dispatch products from ETE and other companies.

The MobileTrak system consists of the following three main components:

○ **MobileTrak Remote.** A compact, self-contained position tracking and reporting system that includes a wireless digital packet radio and a GPS receiver. You can mount a MobileTrak Remote easily in any vehicle. The system connects to a vehicle's power system or can operate for up to 10 hours with rechargeable Nickel-Metal Hydride batteries.

○ **MobileTrak Mapping Diplay Software.** Displays vehicles or other mobile assets equipped with MobileTrak Remotes on a full color, real-time, moving map display. The software has many features, such as vehicle tracking, zooming, panning, address location, heading and velocity labels, and two-way chat messaging. The Mapping Display Software can also remotely control a MobileTrak Remote via commands sent over the wireless data communications network.

○ **MobileTrak Wireless Agent.** A software product supporting wireless communications between the MobileTrak Mapping Display Stations and vehicles or assets equipped with MobileTrak Remotes. The MobileTrak Wireless Agent supports up to 100 MobileTrak Display Stations.

In the United States, the MobileTrak Wireless Agent supports both RAM Mobile Data and ARDIS wireless WAN services. Internationally, the MobileTrak Wireless Agent supports DataTAC and Mobitex wireless networks.

Rockwell's FleetMaster

Rockwell has a product called FleetMaster, which uses CDPD and GPS to provide a vehicle location system. The FleetMaster system includes an in-vehicle Rockwell NavCell unit, as well as a base. The NavCell unit receives and processes signals from the Rockwell's GPS constellation of satellites. Via a wireless WAN link, FleetMaster communicates the time, position, speed, bearing, and status of the vehicle to the dispatch center.

Prince and SkyTel, the automotive electronics integrators, have the product AutoLink System to track automobile locations. AutoLink uses a GPS receiver from Motorola and narrow band PCS services from SkyTel to provide automatic emergency response, theft deterrence, vehicle tracking, two-way personal paging, remote vehicle unlocking, driver personalization, navigational guidance, and location-based information service.

Wireless WAN Case Studies

Guaranteed Overnight Delivery (G.O.D)[1]

G.O.D. is a leading New Jersey-based trucking company employing over 120 drivers servicing the east coast from Virginia to Maine, operating in a competitive industry with the slimmest of margins. Every competitive edge, customer service improvement, and cost reduction accomplishment drives right to the bottom line.

G.O.D. operates a hub and spoke distribution system and collects cargo in a central location prior to long haul shipping to one of the 41 east coast locations. Local drivers take deliveries from these sites to the customers. At the end of the day, drivers collect shipments to be rerouted throughout the central hub and the process repeats itself. G.O.D. drivers typically make 10–20 stops per day, all while reporting problems and their vehicle locations to dispatch centers.

G.O.D. had been working with a two-way communications system. The central office would page the drivers, and the drivers would stop to call headquarters from a pay phone (using cellular phones was judged to be too expensive). In addition to struggling with a two-way communications system that didn't meet G.O.D.'s communications needs, the end-of-the-day workload at the warehouse forced G.O.D. to juggle incoming freight without any real advance notice, resulting in high overtime costs.

RAM Mobile Data brought a complete solution to G.O.D.—a real-time wireless data system that involves Fujitsu's pen based computers equipped with Ericsson wireless modems. The system runs route management software from Roadshow International over RAM's packet radio network.

Now, G.O.D. drivers and the home office are in constant touch—delivery and pickup status, scheduling changes, and route changes are communicated without interrupting the delivery process. Customers get better service, drivers waste less time and make more customer stops during the day, and the warehouse is constantly updated on the loads being brought back at day's end for consolidation and rerouting.

Pepsi Cola-Allied Brothers[2]

Recognizing the need to expedite the sales/order entry process, Pepsi Cola-Allied Brothers looked beyond its existing automated voice-synthesized telephone keypad entry system. As a large franchise bottler serving 35 counties in Maryland, Connecticut, and New York, orders were relayed over a single line, and only one salesperson could enter orders at a given time. Worse yet, the system didn't work with all touch-tone telephones. Route salespeople became frustrated because the telephone line would constantly be busy, and, when the connection was made, many phones were not compatible and valuable time was wasted on the telephone.

Meanwhile, delay would build up. At the warehouse, the distribution personnel would be assigned to stage orders beginning at 11:30 a.m., but the sales orders didn't come in until 1:30 to 4:00 p.m., generating a bottleneck for orders.

Handheld computers from Norand and networking software from Racotek were used in conjunction with RAM Mobile Data's networks to provide a more efficient wireless data sales and order entry process. The combination of these elements and the rising efficiencies in wireless order transmission saves each member of the sales staff one and a half to two hours every day. This savings provides more time for customer service, sales, and merchandising. The end result of using RAM Mobile Data is that Pepsi-Allied's employees in the field and the warehouse are working smarter, not harder. Overtime in the warehouse has been reduced and customer orders arrive on time.

Jacksonville Electric Authority[3]

The propensity of Florida's Jacksonville Electric Authority's (JEA) operating region to experience severe weather conditions forced the company to review its reliance on voice radio communications. Storm force winds, including Hurricane Andrew, caused the loss of wireline service and left up to 80,000 JEA customers without power. These critical periods typically resulted in cellular telephone networks overloading and congesting JEA's voice radio systems as storm repair began.

JEA equipped 38 field service troubleshooters with AST Research Inc.'s GRID 1680 notebook computers, an Ericsson C719 wireless radio modem in conjunction with RAM Mobile Data's service, and TelePartner International Inc.'s Mobi3270 wireless software.

The distribution system restoration process began with the dispatcher taking data from a customer call and generating a repair ticket. Prior to the RAM deployment, the dispatcher alone had access to the information center, and, as a result, had to disseminate the information to the field technicians over the voice radio network. The dispatcher's attempts to contact field technicians were time consuming and wasted the skill of both parties. Storms would only exacerbate the problems.

The RAM Mobile Data system has radically improved the ability of JEA to respond to severe weather conditions. The dispatcher and the field technicians now have access to the same distribution system restoration information. Field technicians can access information simultaneously and the dispatcher is able to spend more time answering customer inquiries and redirecting power around trouble spots.

Field service troubleshooters can access key online repair information, such as customer names, phone numbers, meter numbers, transformer locations, and a detailed outline of the problem. Further, dispatchers can determine the status of repairs and the availability and location of troubleshooters in the field, enabling the dispatcher to respond to evolving weather-network problems.

JEA has not abandoned their voice radio facilities. The implementation of the RAM Mobile Data service reduced the volume demands on the voice radio, enabling it to be held in reserve for more complex repair tasks.

When JEA first implemented the mobile data capability, 66 percent of the repair and maintenance jobs no longer required voice communications between the field technicians and the dispatcher. That figure has since risen to 90 percent as system efficiencies increase.

Athena Rideshare Project Conducts Successful Demonstration of Wireless Navigation Technologies[4]

ETE, Inc. participated in a successful demonstration of TRW's dynamic rideshare project, Athena, which is a wireless, high-technology transportation system that will add personalized transportation options to traditional public transit and carpooling. Athena features a database, advanced wireless communications, and automatic vehicle location (AVL) systems to coordinate rides between providers and riders of many modes of ground transportation, including personal vehicles, buses, vanpools, and taxis. The Athena project is the next step in making vehicles smarter so we can move people safely and conveniently at the lowest possible cost to everyone. The alternatives to driving alone will combine the comfort and convenience of a taxi with the economic and environmental benefits of a carpool.

The goal of the demonstration was to verify the performance of the AVL and communications technology in an actual operating environment. The demonstration was conducted from the offices of L.D. King of Ontario, California, the prime contractor for the $2.2 million Athena project. The demonstration consisted of a rideshare van equipped with an ETE MobileTrak AVL System Remote device and an Apple MessagePad. The van was sent out to roam Ontario's streets while a volunteer went to a randomly selected location within the city to call for a ride. The team watched the van cruise the streets on a computer-generated color display of all the city streets and freeways.

The volunteer called from a street corner, and the location and driving directions were sent to the pick-up van. Within one minute, the van symbol on the map changed directions and proceeded to the caller's location. The AVL system was accurate enough for the team to tell when the van made a wrong turn. That accuracy enabled the team to send directions to the van to turn around and prevent it from getting lost.

With the ETE low-cost hardware components successfully tested, the next step is to decide upon the software that will automatically coordinate ride requests and dispatch up to 100 vehicles to provide ridesharing. The TRW team will present its software design in early 1996, and expects to have a demonstration system operational within one year.

Aegis Transportation Systems of Beaverton, Oregon is supporting operations analysis and TRW is serving as the system designer and integrator. The contract is funded through the FHWA and CalTrans New Technology Division and administered by the City of Ontario.

The Athena team called upon ETE to provide its MobileTrak AVL equipment and wireless software services for this demonstration. ETE's MobileTrak AVL system integrates GPS and wireless data communications. ETE also integrated, under contract, a custom user interface to an Apple Newton Personal Digital Assistant (PDA) for the Athena demonstration. RAM Mobile Data, a business partner with RAM Broadcasting and Bell South, provided the communications services through its nationwide public radio network.

---

[1] Reprinted by permission from RAM Mobile Data
[2] Reprinted by permission from RAM Mobile Data
[3] Reprinted by permission from RAM Mobile Data
[4] Reprinted by permission from ETE, Inc.

PART II

Analyzing the Need for Wireless Networks

Managing a Network Project

Organizations perform work to accomplish their overall business objectives. In some cases, this work might be part of an operation that is continuous and repetitious. As the system administrator for a client server system, for example, you might perform daily back-ups of databases. This task, as well as others, is part of the operation of systems management. Projects are similar to operations; they are performed by people, constrained by limited resources, and should be planned, executed, and controlled. Projects, however, are temporary endeavors people undertake to develop a new service or product. Thus, you can classify network implementations as projects because they have a definite beginning and end.

Before embarking on a networking project, whether the implementation includes wireless networking or not, you should understand the concepts of how to organize a team, plan the activities, and handle risks and contingencies. This chapter defines the following concepts and steps necessary to ensure a successful project:

○ Establishing project management principles

○ Planning the project

○ Executing the project

○ Utilizing project management tools

Establishing Project Management Principles

The Project Management Institute (PMI)[1] defines project management as the art of directing and coordinating human and material resources throughout the life of a project. As shown in figure 5.1, project management primarily consists of planning, monitoring, and controlling the execution of the project. Planning involves identifying project goals and objectives, developing work plans, budgeting, and allocating resources. Project monitoring and control ensure that the execution of the project conforms to the plan by periodically measuring progress and making corrections to the project plan if necessary.

Planning	Monitoring and Control
• Identify Project Goals and Objectives • Develop Workplans • Develop a Budget • Allocate Resources	• Periodically Review Work Progress, Schedules and Budget. • Make Changes to the Project Plan if Necessary

Figure 5.1
The elements of project management.

The use of sound project management principles results in many benefits, such as:

○ Clarification of project goals and activities

○ Better communication among project team members, executives, and the customer

○ Accurate projections of resource requirements

○ Identification and reduction of risks

○ More effective resolution of contingencies

Benefits such as these help an organization complete a quality network implementation on time and within budget.

Planning the Project

Planning is an important part of almost all activities. For example, before you leave for a five-day camping trip in Desolation Valley in a remote area of the Sierra Nevada Mountains, you should make a list of the activities you're going to perform, such as eating and sleeping, and pack the items you will need. Not planning ahead could lead to disaster because you might forget to take something important, such as a flashlight or a water purification kit. As shown in figure 5.2, project planning will enable your project to take the most direct path in reaching project goals.

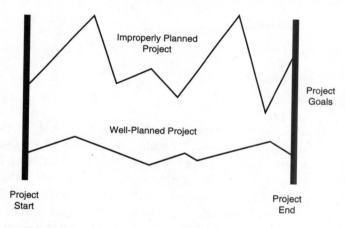

Figure 5.2
The effects of project planning on reaching project goals.

Specifically, project planning is a process consisting of analysis and decisions for the following purposes:

○ Directing the intent of the project

○ Identifying actions, risks, and responsibilities within the project

○ Guiding the ongoing activities of the project

○ Preparing for potential changes

In the planning stage of a project, visualize the goals you have for producing a product or service and then document the thoughts and actions necessary to maximize a successful outcome. In some cases, you will need to determine the requirements and any necessary products before you can complete the project plan.

You should produce a project plan by performing the following steps:

1. Define the project scope

2. Develop a work plan

3. Create a schedule

4. Identify resources

5. Develop a budget

6. Define project operations

7. Evaluate risks

After evaluating risks, you might need to refine some of the other elements of the plan. A project, for example, might require the team to interface a handheld wireless data terminal to an existing IBM mainframe computer containing a centralized database. If the team's design engineer has no experience working with mainframe databases, you should consider the project at risk and attempt to mitigate the problem. Most likely, you would modify the resource plan by either assigning another employee to the project or utilizing a consultant to assist when necessary.

In fact, you should treat the project plan as a "living document"— one that you should update as more information, such as detailed requirements and design, become available.

Identifying Project Scope

Before determining project tasks, staffing, a schedule, and the budget, you must first define the project's scope, which provides a basis for future project decisions. The project scope gives a project team high level direction, allowing an accurate development of remaining planning elements and execution of the project. For each project, you should prepare a project scope having at least the items shown in figure 5.3 and described in the following paragraphs.

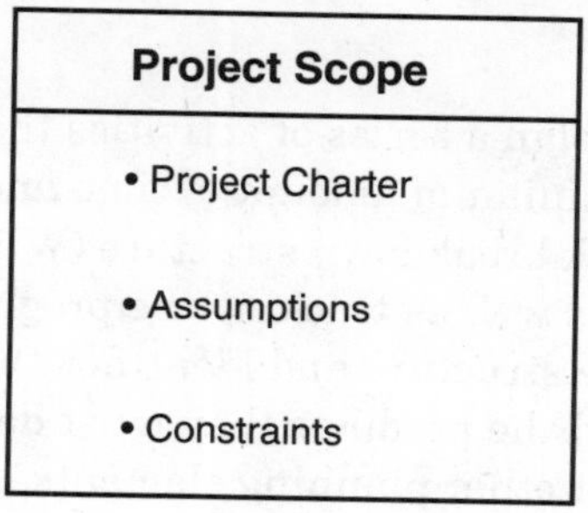

```
┌─────────────────────────────────┐
│         Project Scope           │
├─────────────────────────────────┤
│                                 │
│  • Project Charter              │
│                                 │
│  • Assumptions                  │
│                                 │
│  • Constraints                  │
│                                 │
└─────────────────────────────────┘
```

Figure 5.3
Elements of the project scope.

Project Charter

The *project charter* formally recognizes the existence of the project, identifies the business need that the project is addressing, and gives a general description of the resulting product. The product description defines the main characteristics of the product that the project will create. It should also show the relationship between the product and the business needs of the organization. The requirements phase of the project will define more details of the product. A manager external to the project should issue the charter and name the person who will be the project manager. The project charter should provide the project manager with the authority to apply people and material resources to project activities.

Assumptions

The project team should state *assumptions* for unknown or questionable key factors that could affect the project. A product vendor, for example, might tell you that a new wireless device will be available on a specific date. If the success of the project is dependent on this product, then you should identify its availability as an assumption. This will assist you when evaluating project risks.

Constraints

Constraints limit the project team's options in completing the project. Common constraints are the amount of funding, technical requirements, availability of resources, type and location of project staff, and schedules.

Developing a Work Plan

To reach the goals of the project, plan a series of activities that produce the end product with a minimum amount of time and money. The development of a work breakdown structure (WBS) is a good way of planning the tasks, as well as tracking the progress of the project. A WBS has a tree-like structure and identifies the tasks the team will need to perform and the products they must deliver. The WBS also provides a basis for other planning elements, such as resource allocations and schedules.

As shown in figure 5.4, the first level of the WBS should indicate major phases, followed by lower layers that identify tasks and subtasks. A common question is what level of detail should the WBS include? As a minimum, you should specify enough detail so it is possible to determine the length of time to complete and estimate the cost of each phase and task. This will make it possible to more accurately plan the project.

1. First Major phase
    1.1 First task of phase 1
        1.1.1 First subtask of task 1
          •   •
          •   •
          •   •
          •   •

2. Second major phase
          •   •
          •   •
          •   •
          •   •

Figure 5.4
The organization of a work breakdown structure.

Most networking projects will consist of, in addition to the planning stage, major phases as shown in figure 5.5. Chapters 7–10 focus on how you should complete these phases of the project. For now, the following list gives an overview of each phase.

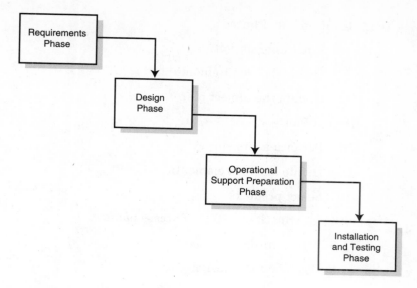

Figure 5.5
The major phases of a networking project.

○ **Requirements Phase**. Defines the organization of the eventual network or system and the needs of its users, providing the basis for a solution.

○ **Design Phase**. Consists of selecting a set of technologies, standards, and products that satisfy requirements.

○ **Operational Support Preparation Phase**. Consists of the planning necessary to effectively support the system after it is installed. Preparations include training development and delivery, and plans for support elements such as maintenance, system administration, and security.

○ **Installation and Testing Phase**. Often referred to as implementation, this phase installs the network components and runs tests to verify proper operation.

The following is an example of a two-level WBS for a typical networking project. In some cases, you will need to identify a set of subtasks for each task.

1. Requirements Phase

    1.1 Elicit information

    1.2 Define the requirements

    1.3 Update the project plan

2. Design Phase

    2.1 Perform a site survey

    2.2 Define network elements

    2.3 Select products

    2.4 Identify the location of access points

    2.5 Verify the design

    2.6 Document the design

    2.7 Obtain approvals for the design

    2.8 Procure components

3. Operational Support Preparation Phase

    3.1 Prepare training courses

    3.2 Define system administration staffing and procedures

    3.3 Establish a help desk

    3.4 Define network management methods and procedures

    3.5 Establish a maintenance process

    3.6 Establish an engineering staff

    3.7 Define configuration control procedures

4. Installation and Testing Phase

    4.1 Plan the installation

    4.2 Install the components

    4.3 Test the installation

    4.4 Transfer the network to operational support

Creating a Schedule

The schedule indicates the element of timing in a project, making it possible for the project manager to coordinate work activities. The schedule and WBS are the basis for selecting and coordinating resources, as well as the primary tools for tracking project performance. A schedule should contain the following information:

○ Names of the phases and tasks listed on the WBS

○ Names of the persons having responsibility for each task

○ Starting date, duration, and due date of each task

○ Relationships between phases and tasks

The project manager should create the schedule by first recording the phase names listed in the WBS and assigning someone to be responsible for each. The next step, working with the responsible team members, is to determine the starting date, duration, and due date for each task. If you cannot determine these characteristics for each task, consider further division of the task into subtasks to accommodate a more accurate assessment. You should also indicate the relationships between tasks using precedence relationships. In other words, show conditions that must be met (such as the completion of a particular task) before starting each task.

A project team must often deal with unrealistic schedules; that is, there might not be enough time to complete a quality implementation. In this case, you might want to consider decreasing the scope of the project.

Identifying Resources

Resources are the people and materials you need to perform the activities identified in the work plan. The goal of resource allocation, like most other planning activities, is to assign people and materials that maximize the success of the project, while minimizing the cost and time to complete the project. As you identify the resources, confirm their availability and schedule them to ensure they are ready when needed.

To properly plan resources, you need to

1. Establish a project team

2. Identify necessary materials

Establishing a Project Team

You should strive to formulate a project team that is fully capable of completing the project tasks. A project team with the goal of implementing a wireless network should consist of a project manager, customer focal point, analysts, engineers, implementors, and operations representative (see fig. 5.6).

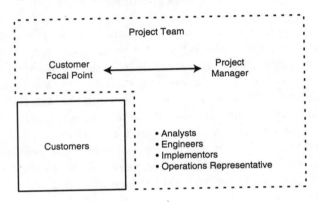

Figure 5.6
The elements of a project team.

○ **Project Manager**. The team should have one project manager who manages, directs, and is ultimately responsible for the entire project. This person coordinates the people and resources, ensuring all objectives of the project are met on time and within budget. The project manager should have experience and education in managing projects, have excellent communications skills, and be familiar with wireless networking concepts and with the customer's environment.

○ **Customer Focal Point**. A single customer focal point represents the interests of the users of the network and aims the project team in the right direction when determining requirements. The customer focal point should be very familiar with the user population and be able to speak honestly for the users.

○ **Analysts**. Analysts gather information and define the needs of the users and the organization. The analyst should have good interviewing skills and be able to translate user and organizational needs into system requirements. It is also beneficial that at least one analyst on the team fully understands the customer's business area.

○ **Engineers**. Engineers provide the technical expertise necessary to fulfill the objectives of the project. Engineers should be part of analyzing needs, but primarily they work on designing solutions that satisfy requirements. Thus, engineers should be very familiar with wireless technologies and should understand how to interface wireless products to existing networks and systems. In addition, engineers can assist with installing the network components.

○ **Implementors**. The implementors are the technicians who install and test the network. The installers set up and interface network hardware, software, and wiring; therefore, they should be familiar with reliable installation practices. Testers ensure the installation meets user expectations, system requirements, and quality standards.

○ **Operations Representative**. The project team should have one operations representative to coordinate the project with existing network support organizations, ensuring that the implementation integrates well into the existing network infrastructure and support mechanisms. Thus, the operations representative should have a good knowledge of the existing network and understand current network support mechanisms.

How many analysts, engineers, and implementors should you have on the team? There are no accurate rules of thumb because the number of staffing depends on the complexity of the customer organization, the scope of the project, schedule constraints, and the experience levels of the people you have available to perform the work. If you are planning to deploy a wireless data entry system for a business having 50 employees, then you can probably get by with one or two team members. A deployment of this system to a company with 5,000 users, however, will require several analysts and engineers to define requirements and design the system, as well as a cadre of installers. The most important thing, though, is to make certain the team is composed of people having the ability to complete the project on time.

Identifying Material Resources

Besides identifying people resources through the development of a project team, you will need to indicate other resources, such as computers and software. Computers and associated software might be necessary to manage the project and create requirements documentation, design specifications, and installation and support plans. Other tools, including scheduling software, CAD (Computer Aided Drawing), and simulation software, might be necessary to effectively complete the project. Figure 5.7 illustrates an outline of a typical resource plan.

```
┌─────────────────────────────────────┐
│           Resource Plan              │
├─────────────────────────────────────┤
│  • Project Team Staffing             │
│  • Office Space Requirements         │
│  • Computer Requirements             │
│  • Software Requirements             │
│                                      │
└─────────────────────────────────────┘
```

Figure 5.7
A resource plan outline.

Developing a Budget

As part of the decision to begin a project, managers might have performed an economic analysis and allocated a specific amount of funding for the project. Thus, the project team might need only to validate and refine the budget, given the knowledge of the work plan and staff availability. If no previous budgeting has been done, the team will need to start from scratch. For this case, estimate hardware and software costs by performing some requirement assessment and preliminary design. Most system integrator companies refer to these as pre-sale activities, providing a basis for a preliminary budget.

The WBS, schedule, and resource plan provide the basis for determining the cost of the project. Before estimating the cost, you will need to assign resources to each WBS task. In other words, identify who will be managing and performing the work for each task and indicate any necessary materials. The next step is to calculate labor and material (including travel) costs for each task, phase, and the entire project. Again, you might need to perform at least a preliminary requirements assessment and design before being able to determine costs associated with the hardware and software of the system being implemented.

During the execution of the project, you will need to track whether the project is being completed within budget. To facilitate budget control, assign unique account codes to project phases and subcodes to each WBS task. During the planning stages of the project, the initial budget is likely to be merely an estimate. After completing the requirements and design stages, the team might need to adjust the budget to reflect more precise information. Figure 5.8 identifies the major items of a project budget.

Project Budget

- Labor Costs
- Hardware and Software Costs
- Travel Costs
- Meeting Expenses

Figure 5.8
The major items of a project budget.

Defining Project Operations

The project scope, work plan, schedule, resources, and budget are the physical makeup of the project. To ensure a project runs smoothly, however, you should also define project operations by developing an operations plan. This plan covers the rules and practices people should follow during the project.

What aspects of the project should the operations plan cover? Generally, you should specify procedures for project actions that need to be followed each time the action is required. For instance, you should include these items in the operations plan:

- Roles and responsibilities of team members
- Methods for coordinating with other organizations
- Staffing procedures
- Travel policies
- Engineering drawing standards
- Document sign off procedures

To develop a project operations plan, review existing corporate and local policies and regulations, then define procedures for items that team members must accomplish in a unique manner. Also, be sure to identify any restrictions that corporate policies and other regulations place on the project.

Dealing with Project Risks

The success of a project is often jeopardized by unforeseen elements that crop up at inopportune times. The nasty truth is that many projects are not completed on time, within schedule, or as expected. As an example, a project team might successfully complete the design stage of the project and be ready to purchase the components when they discover the customer's upper management has lost interest in the project and has withheld further funding. Or, you do a thorough job of defining user needs and then the team is not successful at determining a set of technologies and products that will fulfill the requirements.

To maximize the success of a project, the project team must not only develop a WBS, schedule, and resource plan, but also continually identify and manage risks. Risk management should begin early in the project, even during the planning stage, then continue throughout the project. A risk factor usually has more impact if you don't attempt counter measures until later in the project. To avoid negative consequences, the team can manage risks by identifying risk factors and determining methods to reduce them. A risk factor is anything that might have adverse effects on the outcome of the project.

Here are examples of risks factors you should consider:

- Project Factors
  - Clarity of project objectives
  - Project team size
  - Working relationships among project team members
  - Team geographical disbursement
  - Project duration
  - Delivery date constraints
  - Project manager's prior experience

- Resource Factors
  - Experience of project team members
  - Team learning curve
  - Use of contractors
  - Potential loss of team members due to other projects

- Organizational Risk Factors
  - Level of management commitment
  - Funding constraints
  - Level of user involvement and support during the project
  - Firmness of benefits
  - Length of time necessary to receive a return on investment for implementing the project

- Technical Factors
  - Range of technologies and products available to satisfy requirements
  - Complexity of the interfaces to existing systems
  - Familiarity and prior experience with the type of implementation the team is accomplishing

You can control risks by following these steps:

1. Review the project's WBS, schedule, resource plan, and budget and assess the status of the preceding potential risk factors.

2. Define the potential impact each risk has on the successful completion of the project.

3. Pinpoint the causes of the risks.

4. Refine the work plans to reflect the risk reduction strategy.

5. Periodically reevaluate the potential risk factors, especially those found earlier in the project, and take necessary counter-active measures.

Creating the Project Plan

Before placing the project in full motion, you should collect all the individual planning elements and create a project plan as shown in figure 5.9. This could be as simple as putting a cover around the planning documents you already prepared. You can use the project plan to effectively manage the project through all phases.

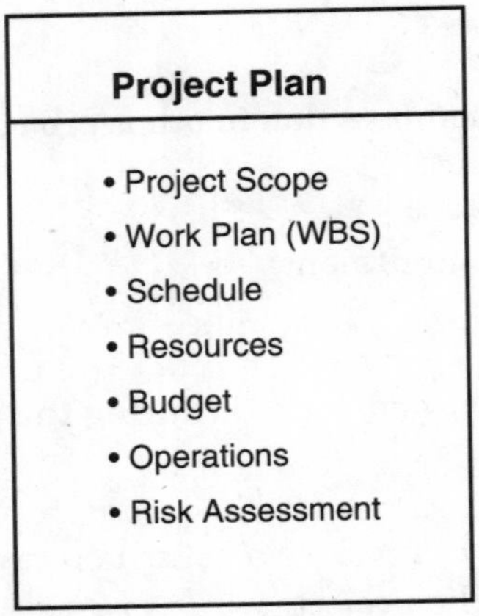

Project Plan

- Project Scope
- Work Plan (WBS)
- Schedule
- Resources
- Budget
- Operations
- Risk Assessment

Figure 5.9
The project plan.

Most organizations require a review and approval of the project plan by upper management before executing the project. As a final check before giving your project plan to an executive, use the following list to be certain the project plan:

○ Has a project scope identifying clear objectives and business needs.

○ Contains a work breakdown structure that covers all activities necessary to accomplish the project's objectives.

○ Contains a work breakdown structure that is detailed enough to accurately define the schedule and budget.

○ Spells out a schedule that is realistic.

○ Identifies a project team with team members who are available when needed and have the experience necessary to successfully complete the project.

○ Identifies needed material resources, such as software tools and travel.

○ Contains a budget with accurate costs.

○ Adequately defines the roles and responsibilities of team members.

○ Defines all risk factors that might impact the project, as well as actions needed to counteract the risks.

Executing the Project

After completing the planning stage of the project, the project manager can begin work activities with a kick-off meeting and guide the project through the activities identified in the WBS. The project team should periodically hold status meetings to assess the progress to date and make changes to the plan if necessary to keep the project on course. Figure 5.10 illustrates these project management actions.

Figure 5.10
Project management activities.

The Kick-Off Meeting

The entire project team should have a kick-off meeting to review the project plan and officially start the project. This starts the team off together and avoids having people stray away from the primary objectives. Figure 5.11 is an agenda you should use as a basis for the kick-off meeting.

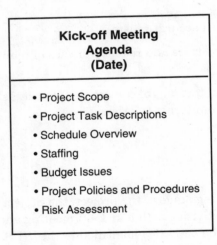

Figure 5.11
A Kick-off meeting agenda.

The key to an effective kick-off meeting (or any meeting for that matter) is to stay focused by keeping discussions within scope of the specific agenda items. Spend a few minutes at the beginning of the meeting to review the agenda and ensure everyone agrees the topics are applicable and see if anything is missing. It's not too late at this time to make alterations to the agenda if necessary.

Periodic Activities

Periodically, the team should check the status of the project, perform technical interchange meetings, and report progress to upper management. The following list explains each of these activities:

○ **Status Checks**. For most projects, a weekly or biweekly status check is often enough to review project progress. You can normally accomplish this at a project staff meeting. The project manager should at least review completed tasks and check whether the project is on time and within budget. It's also not a bad idea to review risk factors and take action to minimize their impact.

○ **Technical Interchange Meetings (TIMs)**. TIMs address technical issues needing attention by project team members and customer representatives. A TIM is effective if the solution to a technical requirement or problem cannot be adequately solved by a single team member. In this case, schedule a TIM and invite the people needed to crack the problem.

○ **Progress Reports**. Progress reports summarize the technical, schedule, and cost status of the project. The main idea is to show a comparison between planned and actual elements. Project managers should periodically send these progress reports to upper management to keep them abreast of the status of the network development. It is normally best to alert management of conditions that might impact the project as early as possible. This allows enough time for upper management to assist in countering the problems. Also, be sure to include tasks the project team still needs to complete, especially the ones that are planned to take place up until the next progress report. Figure 5.12 identifies the key components of a progress report.

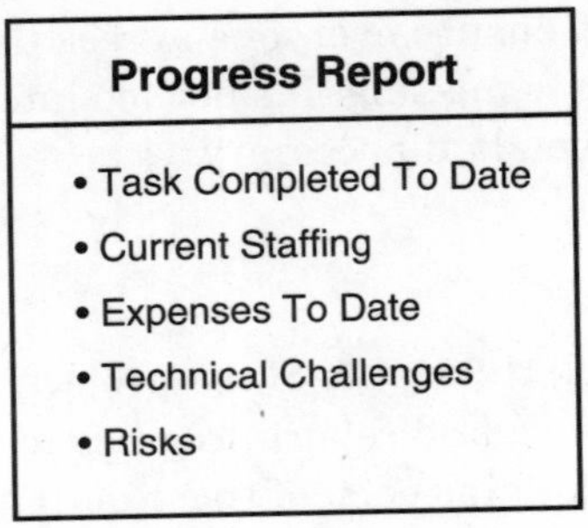

```
┌─────────────────────────────────┐
│         Progress Report         │
├─────────────────────────────────┤
│  • Task Completed To Date       │
│  • Current Staffing             │
│  • Expenses To Date             │
│  • Technical Challenges         │
│  • Risks                        │
└─────────────────────────────────┘
```

Figure 5.12
The key components of a progress report.

A management report should focus on current accumulative costs and the schedule status, past and present resource utilization, negative impacts on the project schedule, identification of successful and unsuccessful tasks, as well as major changes made to the project plan. Major changes to the project plan should also be thoroughly explained. The progress report also should explain how the project team will counter all deficiencies.

Maximizing Communications

During the execution of the project, take steps to maximize communications flow among team members when determining requirements, designing the system, and performing installations. The problem with many project organizations is that they operate in a very serial form as shown in figure 5.13. As a result, they depend heavily on documentation to convey requirements, solutions, and ideas. In this case, the customer represents the needs of potential end users of the system or product under development. In companies that develop software products, sales and marketing staff typically express customer needs in terms of requests and requirements. Otherwise, requirements generally flow directly from the customer. Project managers are often responsible for managing the overall development, installation, and support of the product or system. Typically they produce the first specification the development group uses to design and code the system or product.

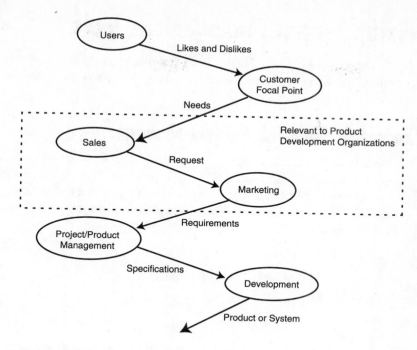

Figure 5.13

Serial communications flow through an organization.

There are several problems with this process, which lead to systems and products that do not adequately meet the users' needs. The series of hand-offs between the different players in the process, for example, can take a long time, delaying the creation of a prototype for validation purposes. In addition, the process doesn't engage the customer continually throughout the process, forcing developers to guess at missing or incomplete requirements. The process also dilutes the clarity of requirements as they flow via documentation and the spoken word from one element to the next.

The solution to this serial communications problem is to utilize team meetings that incorporate representatives from all organizational groups, especially when defining requirements. Sometimes this is referred to as Joint Application Design (JAD). This concept is described in Chapter 6, "Defining Requirements for Wireless Networks."

Utilizing Project Management Tools

There are many tools available on the market that can assist in planning and executing the project. Here are a few software products you should consider.

Primavera's SureTrak Project Manager

SureTrak Project Manager is ideal for resource planning and controlling on small to medium-sized projects. With SureTrak, you can create schedules graphically and point and click to create relationships between activities, simplifying updates. SureTrak offers the following features:

- Project Modeling
  - 10,000 activities per project
  - Scheduling and resource leveling
  - Customizable activity and resource calendars
  - Target date, original/remaining duration, and budget baseline comparisons
  - What-if analysis
  - Earned value analysis
  - 20-level Work Breakdown Structure
  - 10-level outlines
  - 24 activity codes with dictionary of titles
  - Use activity codes and IDs for selection, sorting, grouping, summarizing, and reporting.
- Scheduling
  - Critical path scheduling
  - Precedence Diagramming Method (PDM)
  - Finish-to start, start-to start, and finish-to-finish relationships
  - All relationships with lead and lag

- Free, total and negative float calculations
- 31 activity calendars per project
- Durations in hours, days, weeks or a combination
- 10 types of schedule constraints (for example, start-no-earlier-than, finish-no-later-than, as late as possible)
- Suspend and resume dates for in-progress activities
- Automatic and manual activity updating

- Resource and Cost Management
  - Unlimited resource calendars
  - Resource leveling with customizable priorities
  - Resource-driven durations
  - Unit cost and revenue by resource
  - Variable resource availability
  - Resource start and finish dates and lags
  - Cash-flow forecasting
  - Track budget, actual cost to date, percentage complete, earned value, cost to complete, cost at completion, revenue to date, revenue to complete, and revenue at completion
  - Cost, schedule, and budget variances

- Presentations, Reports, and Graphics
  - Unlimited presentation style layouts
  - Reuse layouts with any project
  - WYSIWYG editing
  - Complete customization of project workspace (for example, colors, date, and time formats)
  - Time-scaled bar (Gantt) chart
  - Time-scaled logic chart
  - Tabular/columnar view

- Resource/cost histograms and curves

- Resource loading tables

- User-definable start and end-points, colors, fonts, sizes, positions

- Print to fit (to specified number of pages or scaling percentage)

- Organize by any combination of activity codes, dates, and resources

- More than 40 predefined reports and graphics

- Customizable header and footer with titles, dates, logos, legends, and revision blocks

- Attach pictures or text to bar chart

- More than 30 predefined layouts

- Multiple levels of selection (filter) criteria

- Predefined and customizable filters

- Custom report writer

- Summary and discrete activity bars

Primavera's Project Planner

Primavera's Project Planner (P3) enables you to efficiently manage large, multiple, complex projects. It offers a single database solution (ODBC-compliant) that provides simultaneous access to project files by multiple users throughout the company.

The following are P3's main features:

- **Control**. With P3, you can create a project group to reflect complex links among individual projects. P3 lets you define any number of interrelated master projects and subprojects. The P3 interproject relationship manager lets you link activities between separate master projects, even remote ones. It's an ideal tool for coordinating independent efforts in multiple locations.

○ **Scheduling**. P3 offers powerful scheduling and resource leveling options, including backward resource leveling, smoothing, and the ability to level selected portions of a project. User-definable options help you calculate float and lag in different ways during scheduling and leveling. P3 shows you the effect of progress on remaining work. You can even schedule or level by subproject independently of the master project.

○ **Multiuser**. Users can simultaneously update, analyze, and report on the same project because P3 uses individual record locking, providing maximum concurrent access to your data. Security is assured: P3 can restrict access to certain project data by task, resource, phase, or user-defined filters. Project managers can assign read/write, read-only, exclusive-access or no-access on a master/subproject or a subproject level. For updates from local or remote teams, P3 lets you communicate via Microsoft Mail, cc:Mail, or any other VIM or MAPI-compliant mail system.

Primavera's Monte Carlo

Monte Carlo 3.0 from Primavera is a new Windows product that helps P3 users create better, more realistic schedules and resource plans. Monte Carlo takes P3 schedules and simulates project performance to assess the likelihood of finishing on time and within budget. Users can evaluate the whole project or individual segments based on a quantifiable measure of risk. Monte Carlo is an analysis tool that helps users make decisions, develop contingency plans, evaluate mitigation strategies, and manage risk.

Microsoft's Project

Microsoft's Project is an effective tool for managing projects of various sizes. Project is feature-rich and contains planning wizards that monitor your actions and offer suggestions that can automatically be incorporated into your plan.

Project offers the following features:

- ○ **Critical Path**. Identifies the tasks that risk your project if dates are missed.

- ○ **Drag and Drop Data**. Enter and schedule tasks, assign resources, and establish task links quickly with your mouse.

- ○ **Filters**. View a subset of tasks in your plan, such as project milestones or tasks assigned to a specific person.

- ○ **Baseline Chart**. Compare your current schedule against your original plan.

- ○ **Autocorrect**. Misspellings and typos can be corrected automatically as you type. Or, create your own list of abbreviations that expand automatically.

- ○ **Resource Tools**. Spot bottlenecks and shortages in advance and balance workloads across tasks or projects.

- ○ **OfficeLinks**. Move data easily among spreadsheets, slides, reports, and other documents created with programs in the Microsoft Office family, making information about your project more accessible for your everyday needs.

- ○ **Save to Database**. Directly exchange data with a Microsoft Access database, Microsoft BackOffice, and other databases compliant with the Open Database Connectivity (ODBC) standard.

---

[1] PMI offers a certification titled Project Management Professional (PMP) that you can earn through work experience, education, and successful completion of the PMP examination. The PMP certification ensures you've mastered the skills necessary to manage a project of any type. Many corporations are beginning to recognize the importance of PMP-certified professionals.

Defining Requirements for Wireless Networks

Requirements are crucial in all development projects—they provide the basis for design, implementation, and support of the system or product (see fig. 6.1). Incomplete or missing requirements are the major reasons for unsuccessful projects, resulting in 60–80 percent of system defects that eventually surface late in the development phase or after delivery to the users. These system defects are very time-consuming and expensive to correct. Shabby up-front requirements also lead to the continual stream of "new" requirements that fill in for inadequacies throughout the project. New requirements cause a great deal of rework, extending development time and costs. Requirements that are ambiguous, untestable, and most of all, not able to fully satisfy needs of potential users contribute to high development costs, lagging time-to-market, and unhappy users and customers. Thus, organizations must emphasize the definition of requirements to keep their heads above water.

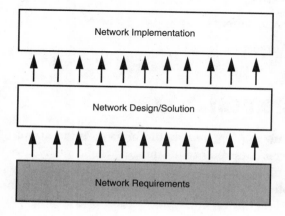

Figure 6.1

Requirements—the foundation of design, implementation, and support.

A large part of planning the requirements phase is to calculate how much work will be necessary to determine what the requirements are. For all projects, thoroughly assess requirements for security and interfacing to existing systems; keep in mind, however, that the amount of time spent analyzing the user's requirements differs depending on whether you are developing software applications or using off-the-shelf products. You should spend more time defining user functional requirements for an application development than for the deployment of just the network infrastructure. The reason is that users spend more time interfacing directly with applications rather than the network components.

For example, a project meant to create a specific graphical user interface (GUI) for nurses to access healthcare records from a centralized database would require a detailed analysis of the business processes the nurses' work involves. A project that creates a wireless network infrastructure for off-the-shelf applications, however, would not need such detail. In that case, you only need to examine high-level business processes and product requirements in order to select the correct wireless components.

Although the system analyst is generally responsible for specifying requirements, the rest of the project team should be involved as well. This chapter describes the types of requirements you need to define and discusses the following steps that the project team should take when defining requirements:

○ Elicit information

○ Define the requirements

○ Update the project plan

Types of Requirements

Before eliciting information, you should have an understanding of the type of requirements you are attempting to define. Figure 6.2 lists common requirement types. Knowing the requirement types helps you focus on gathering the best information related to user needs and system requirements.

```
┌─────────────────────────────────────────────┐
│          Requirements Types                  │
├─────────────────────────────────────────────┤
│                                               │
│   ● User Profile and Interface Requirements   │
│   ● Functional Requirements                   │
│   ● Application Requirements                   │
│   ● Information Flow Requirements             │
│   ● Mobility Requirements                      │
│   ● Performance Requirements                   │
│   ● Security Requirements                      │
│   ● System Interface Requirements             │
│   ● Environmental Requirements                 │
│   ● Operational Support Requirements           │
│   ● Regulation Requirements                    │
│   ● Budget Requirements                        │
│   ● Schedule Requirements                      │
│                                               │
└─────────────────────────────────────────────┘
```

Figure 6.2
Common types of requirements.

User Profile and Interface Requirements

The user profile requirement identifies the attributes of each person
who will be utilizing the system, providing human factors that
designers can use to better select or develop applications. A person's
in-depth experience with Windows-based applications, for example,
would prompt the design team to procure or develop standard
Windows applications. In addition, the profile assists the installation team when assigning user names. The user profile should
identify the person's name, job title and description, level of networking experience, and knowledge of applications. Users will
require some form of interface to the system's databases and other
resources. Most interfaces today are graphical client-server–based
applications that run on Microsoft Windows. The user interface
requirement should indicate screen layouts and leveling of menus.

Functional Requirements

Functional requirements describe what the users and the organization expect the system to do. Therefore, functional requirements run parallel to the tasks and actions users perform. For example, the need to enter packing slip information from all received shipments into a database is a functional requirement. In most cases, application software implements functional requirements.

Application Requirements

Application requirements specify that the system can utilize specific applications. For instance, an organization that uses Microsoft Office for word processing and spreadsheets should indicate that standard as an application requirement. An organization moving from paper-based inventory control to a centralized computer system, however, would probably delay the selection of the actual application until the beginning of the design phase (after all other requirements are known).

Information Flow Requirements

Business processes within companies depend heavily on communications. To complete their tasks, people need to communicate with other people and systems. Because the network's primary role is the support of communication, you must fully define information flow requirements. For this requirement, specify the information path flow between people and systems, types and formats of information sent, frequency of information transmission, and maximum allowable error rates. These requirements will provide a basis for the selection of network components, such as the network interface card, medium, and so forth.

Mobility Requirements

Mobility requirements describe the movement of the users when performing their tasks. Mobility requirements should distinguish whether the degree of movement is continuous or periodic.

When the user or network component must have the capability to utilize network resources while physically moving, they are said to be in *continuous* movement. Examples of users requiring access to network resources while continuously moving include emergency vehicles, military personnel on a battleground, delivery services, and healthcare professionals.

Periodic mobility—often referred to as *portability*—implies the utilization of network resources from temporary locations, but not necessarily while the user is in transit between locations. Portability implies a temporary connection to the network from a stationary point, but the interface associated with a portable connection should be easy to move, setup, and dismantle. Examples of users requiring portable interfaces include point-of-sale cashiers, conference organizers, and employees working from a temporary office facility. When specifying mobility requirements, be sure to identify the users needing mobility and the range of movement each user or component needs.

Performance Requirements

Performance indicates how well a network provides applications and services. You never hear people complain when performance is too high. Low performance, however, creates disgruntlement because users cannot do their work as quickly as they want or are accustomed. For performance requirements, identify expected values for reliability, availability, and delay, as follows:

○ Reliability is the length of time a system or component will operate without disruption. Most product vendors refer to this as Mean Time Before Failure (MTBF).

○ Availability defines the period of time the system must be operational. As an example, the availability could indicate that a network should be operational twelve hours a day from 6:00 a.m. until 6:00 p.m.

○ Delay is the length of time users or systems can wait for the delivery of a particular service.

Security Requirements

Security requirements identify the information and systems that require protection from particular threats. The degree of security depends on the severity of the consequences the organization would face if the system were damaged or if data were lost. Of course, military and law enforcement agencies require high-level security. Security requirements should address the sensitivity of information processed on the network, the organization's security regulations, and probability of disasters, such as equipment failure, power failure, viruses, and fire.

System Interface Requirements

Most likely, the system being developed will have to interface and interoperate with existing systems, such as networks and databases. Therefore, system interface requirements describe the architectures of these systems and the hardware, software, and protocols necessary for proper interfacing. If the interfacing method is not known, then you will need to determine a solution during the design phase.

Environmental Requirements

Environmental requirements state conditions, such as weather, pollution, presence and intensity of electromagnetic waves, building construction, and floor space that could affect the operation of the system.

Operational Support Requirements

Operational support requirements define the elements needed to integrate the system into the existing operational support infrastructure. For example, you should require the inclusion of Simple Network Management Protocol (SNMP) if current network monitoring stations require SNMP.

Regulation Requirements

Some organizations might have to conform to certain local, state, or federal regulations; therefore, be certain to specify these conditions as requirements. Regulations imposing safety and environmental procedures place definite requirements on network implementations. The operation of a wireless radio wave adapter, for example, must conform to Federal Communications Commission regulations. Another example is the use of radio-based wireless products on military installations within the United States. The military's use of these devices is regulated by a special frequency management organization, not the FCC. Therefore, radio-based implementations on military bases must conform to the military's frequency management policies. In addition, the company itself might have policies and procedures, such as strategic plans and cabling standards, that the implementation should follow.

Budget Requirements

An organization might have a certain amount of money to spend on the system implementation. Budget constraints can affect the choice of solution because some technologies cost more than others to implement. The budget requirements should consider the funding plan for the installation project—the availability of funds at specific times. The reason for this is to best plan the procurement of components and scheduling of resources.

Schedule Requirements

Schedule requirements should include any definite schedule demands that will affect the project. By their nature, organizations impose scheduling conditions on projects, such as availability of project funds, urgency to begin a return on investment, availability of project team members, and interdependency between this project and other projects. Define schedule requirements so the team knows the time frames it can work within. For instance, the design team may have a choice of using a current wireless adapter or waiting eight months for the next, faster release. If the organization must have the system operational within three months, then the team would have to choose the existing product.

Eliciting Information

The objective of eliciting information is to gather as many facts as you can relating to each of the requirements types. This information will enable you to define each of the requirements during a later step. The following is a checklist of items you should consider performing when eliciting requirements:

○ Review user needs

○ Review existing systems

○ Review the environment

The following sections explain each of these steps.

Reviewing User Needs

It's a good idea to determine what users' needs are before deploying a system. These needs will lead you to the definition of these types of requirements: user profile and interface, applications, information flow, mobility, performance, security, regulations, budget, and schedule.

The most effective method in reviewing needs of potential users is to interview members of the organization. Who in the organization do you interview? Users and managers. It's generally not practical to interview every user—just a cross section will do. Talk with managers to obtain a high-level overview of what the people do and how they interact with other organizations. During the interview, ask questions that enable you to define specific requirements. The following sections give samples of good questions and describe the requirements you need to define.

> **NOTE**
>
> In some cases, reviewing needs of users will identify weaknesses in the current business processes, motivating business process reengineering, which is a realignment of the way a company operates. In fact, the introduction of wireless networking makes it possible to redesign the current paper-intensive methods to a more mobile and electronic form.

Interviewing Techniques

The interviews should determine the organization's structure, departmental missions, work flow, user profiles, and existing system attributes. Before conducting an interview with a particular set of users, be sure to get permission from their manager. Having the boss's buy-in to the project can result in a better response from the interviewees.

Ask managers the questions relating to the mission and major functions of the staff. Questions you ask the staff should be more specific, relating mainly to the individual. However, be certain you interview a truly representative group of users and don't miss any unique needs.

A day or two before the interview, draft a set of questions and distribute them to give the users time to prepare answers. The following is a sample set of questions:

Questions for Managers

○ What are the functions and major activities of your staff?

○ How does your organization interface with other organizations within your company?

○ What is your current staffing level? What is the projected level?

○ Is your staff's work environment under constant construction?

○ Does your organization rearrange desks frequently?

○ What are your staff's needs for a new system (applications and network)?

○ What security policies does your organization follow?

○ What funds are available for this implementation?

○ What schedule constraints exist for this project?

○ What local, state, and federal regulations exist that might influence the project?

Also, if the manager is to be a system user, ask her the following questions as well.

Questions for Users

○ What tasks do you perform? How do you accomplish these tasks? How do your tasks interface with tasks that other people perform?

○ What mobility do you require when performing your day-to-day activities?

○ Do you travel? If yes, how often do you travel and where do you travel to?

○ What internal and external systems to your organization do you need to access?

○ With which internal and external people do you need to communicate?

○ Which types (data, voice, imagery, video), formats (DOC, PIF, GIF, JPEG), and methods (FTP, e-mail, postal mail) do you use to exchange information?

○ What type of computer do you use? Desktop or portable? CPU type, amount of RAM, size of hard drive?

○ What applications are you currently utilizing? Where do these applications reside (desktop, server, mainframe)?

○ What is your experience using client-server applications?

○ What systems (hardware and software) are currently supporting the applications you utilize?

○ What system availability do you require to perform your tasks?

○ What are your needs for information security?

○ What are your needs for a new system (applications or network)?

○ What schedule constraints exist for this project?

Be sure to schedule an appointment with the potential user and arrive on time. People are busy—you don't want to waste their time or be inconsiderate.

TIP

If possible, have two interviewers during the interview: one to ask questions and the other to take notes. This ensures the capturing of all comments. A recording device, such as an audio tape recorder, can prove beneficial; however, they can intimidate some interviewees.

Written Surveys

A written survey is another method for gathering user needs. The process is as follows: write a series of questions that probe the potential user for information that will enable you to assess specific needs, distribute the survey via mail, and insist that people complete the surveys and return them. Unfortunately, this process often doesn't work as expected. It is extremely difficult and time-consuming to write questions that elicit usable responses. Also, many people will not complete the survey; typical return rates on written surveys are 10–15 percent. As a result, you should stick with personal interviews.

Defining the Business Processes

After you gather information from the managers and users, you should define the business processes; that is, document the function of each organization, the tasks each user or group of users perform, and the types of information people and groups exchange. This process assists you when defining the information flow requirements. As mentioned previously, the level of detail of gathering information depends on whether the project is developing an application or using strictly off-the-shelf products.

Reviewing Existing Systems

User needs are only part of the requirements—existing systems also portray important requirements. Reviewing existing systems helps you to define the system interface and operational support. To review existing systems, begin by interviewing the Corporate Information System (CIS) managers and review system documentation, as discussed in the following sections.

Interviewing Corporate Information Systems Managers

Corporate Information Systems managers, including people in charge of applications development, system implementation, and the mainframe data center, are the best sources of information about the existing systems. Again, interviewing is the best method to use. You should follow the same recommendations described for interviewing users, but your questions should focus more on the technical information.

Following are sample questions for interviewing CIS managers:

- What networks and systems (hardware and software) do CIS currently support?

- What is the current telecommunications/WAN topology? (That is, where are the facility locations and are they interconnected via telecommunications or WAN services?)

- What are the corporate inter- and intratelecommunications links?

- Which documents describe your existing networks and systems?

- What networks and system links are planned for the future?

- What tools are currently used for systems and network management?

- What are the company's requirements for information security?

- What functionality do you envision the company's information system providing in the future? How does this differ from the vision of users and executive management?

- What are the company's business plans that might affect the network architecture? (That is, future staffing, geographical coverage, and so forth.)

- What is the company's budget to deploy this system?

- Are any documents available for review that describe user and organization requirements for information systems?

- What policies do CIS have for deploying networks and systems?

Also, if the manager will be a user of the system, be sure to ask the questions for users.

Reviewing System Documentation

When determining requirements, the project team should review current documentation that provides an accurate description of existing systems. For instance, review the concept of operations to examine system-level functionality, operational environment, and implementation priorities for an organization's information system. Also, review the strategic information system plan, which provides a long-term vision and the general procedures necessary to manage the efficient evolution of the corporate information system. This review provides policies and standards the design team may need to follow. In addition, the organization may have other plans, such as business and employee projections, that the team can consider. Business plans describe the future markets and strategies the company wishes to pursue and are useful when determining the types of applications and services the users might require.

Reviewing the Environment

To determine environmental requirements, consider the conditions in which the network will operate. Gather information by interviewing the company's facility manager and visually inspecting the area.

Here are questions you will need to answer for wireless network implementations:

○ What is the physical building made of?

○ What devices in the area might cause interference?

○ Are there any trees that might block the transmission of line-of-sight radio waves?

○ Does the area occasionally experience severe snow, rain, fog, or smog?

○ Where is it possible to install access points within the building?

○ Where is it possible to install directional antennas on top of the building?

The obvious unseen hindrance to a radio-based wireless network is interference. Thus, in addition to talking with the facility manager about potential interference, consider using a radio-based site survey tool to evaluate radio wave activity within the part of the radio spectrum your components will operate. Most wireless LAN vendors include these site survey tools with their products. You can use a spectrum analyzer to measure the amplitude of signals at various frequencies.

If your project also includes wired network components and systems, perform the following activities to determine environmental requirements:

○ **Investigate the capability to run cables throughout the facility.** Be sure to check above the ceilings to determine whether there is enough room to run the cabling, as well as locate and assess all vertical cabling conduits.

○ **Evaluate the electrical system.** An electrical evaluation provides information on whether the building's electrical system can support the network components. Check to see if adequate building power for the new components is available and whether the building has experienced power outages. If problems with the electrical system exist, recommend appropriate corrective action.

○ **Investigate server and communications room locations.** If applicable, determine locations for the system servers and network hubs. Be certain the room(s) has adequate power, air-conditioning, and space for future expansion.

Defining Requirements

After gathering information, you're ready to define the requirements that will provide the basis for the design. To define the requirements, perform these steps:

○ Determine potential requirements

○ Validate and verify the requirements

○ Baseline the requirements

Determining Potential Requirements

The first step in defining requirements is to identify potential requirements by using the information gathered during interviews, review of documents, and inspections. You can accomplish this by:

○ Conducting a Joint Application Design meeting

○ Assessing constraints

○ Documenting requirements

Conducting a Joint Application Design Meeting

An effective method for drafting requirements is to conduct a series of team meetings using Joint Application Design (JAD) techniques. With JAD, all the active participants work together in the creation of requirements. As illustrated in figure 6.3, JAD is a parallel process, simultaneously defining requirements in the eyes of the customer, users, sales, marketing, project managers, analysts, and engineers. You can utilize the members of this team to define requirements.

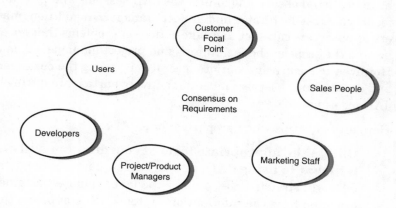

Figure 6.3
Joint Application Design—a parallel process.

The goal of JAD is to reach a consensus on requirements among all team members, especially the customer and developers. JAD ensures the early definition of accurate requirements, minimizing later rework.

JAD is extremely effective for defining requirements because the customer becomes a partner in the development project, allowing an effective customer-developer team, which breaks down communications barriers and increases levels of trust and confidence. Because JAD helps you to determine requirements quickly, developers can start prototyping earlier. Prototyping is important because it provides a vision of the system for the users, fueling the refinement of requirements. JAD also keeps the customer accurately informed on what can and can't be done because engineers can validate the requirements as the customer states them.

In addition to the active participants, JAD consists of a facilitator, scribe, and optional observers (see fig. 6.4). The facilitator manages the overall meeting, acting as a mediator and guide to guarantee the group stays focused on objectives and follows all JAD meeting rules. The facilitator should have good communication skills, be impartial to the project, have team building experience and leadership skills, be flexible, and be an active listener. The scribe records the proceedings of the JAD and should have good recording skills and some knowledge of the subject matter. It may be beneficial to have impartial observers monitor the JAD sessions and provide feedback to the facilitator and project manager. In addition, managers as observers can spot and take action on problems that go beyond the scope of the facilitator's and project manager's domain. However, to ensure appropriate interaction among the customer and developers, observers must not actively participate during the JAD meeting.

Here are some tips when preparing for a JAD:

○ **Obtain the appropriate level of coordination and commitment to using JAD.** In many cases, participation in a JAD will stretch across organizational boundaries. Engineers are often from the Information Systems (IS) group, and the customer might represent users spanning several functional groups. Without concurrence of all group managers, the JAD meetings will appear biased to those not buying into the idea, causing some people to not participate or accept the outcomes. Therefore, to receive commitment to the method, the initiators of the JAD should discuss the benefits and purpose of the JAD with applicable managers of each group.

○ **Ensure there are clear objectives for the JAD meeting.** Without clear objectives, the JAD proceedings will flounder and lead to unnecessary outcomes.

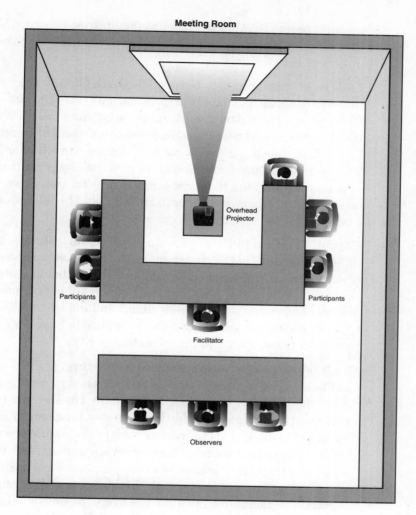

Figure 6.4
Joint Application Design meeting elements.

○ **Consider using an independent consultant as a facilitator.** This assures neutrality and avoids siding with one particular group. Be certain, though, to avert the selection of a consultant closely allied with the department responsible for development. A close alliance could tempt the facilitator to favor the engineers, letting them dominate the meeting and hamper ideas from the customer. It doesn't hurt to have internal people in mind to groom as facilitators; however, be sure they have proper training and are not connected to the project they're facilitating.

○ **Talk to all participants before the JAD.** Discuss the issues involved with the particular project, such as potential conflicts. Give all new participants an orientation to JAD if it's their first time attending one. In some cases, the JAD might be the first time business people and engineers work together. Therefore, minimize communication problems by preparing participants to speak the same language. Avoid using computer jargon. Otherwise, communication could be difficult and customer participation will decline.

○ **Establish rules.** Rules are absolutely necessary because the different agendas of the customer, users, and developers can often derail the JAD and raise conflicts. Rules should state that all members will conform to an agenda, all participants are equal, observers will remain silent, and the bottom line is to reach consensus on requirements. Be sure to have the participants contribute to the formation of rules.

○ **Don't let developers dominate the meeting.** Many JADs tend to have too many developers and not enough representation of the potential end users. This usually blocks users from expressing their ideas. In addition, there's a tendency of IS departments using JAD to "rubber stamp" the requirements— that is, to have the customer merely review and approve them. You should limit the developers to architects and engineers because programmers might push the team toward a design too soon. The facilitator must ensure everyone has fair time to voice their ideas.

Assessing Constraints

As part of the requirements definition, you should identify which of the requirements are constraints. These are firm requirements (ones that can't be easily changed) that limit the choice of solution alternatives. Figure 6.5 illustrates this concept. *Constraints* are usually requirements dealing with money, regulations, environment, existing systems, and culture. However, any requirement could be a constraint if that requirement is absolutely necessary and not subject to change. Regulations are constraints because they often carry a mandate directing a particular form of conformance. The environment, such as building size and construction, establishes constraints because the facility may be too expensive to change to accommodate certain solutions. Existing systems are not always easy to change; therefore, solutions will have to conform to particular platform constraints, memory, and so on.

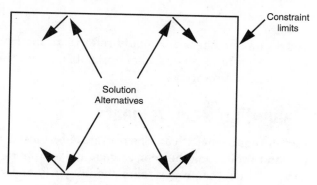

Figure 6.5
The effect of constraints on solution alternatives.

Documenting Requirements

To adequately support the remaining phases of the project, be sure to clearly document the requirements. Without good documentation, requirements can become unclear as time passes and memories lapse. The handover of project information from person to person can also dilute original intentions. To make matters worse, the analysts responsible for defining the requirements could leave and be unavailable during the design phase. Undocumented requirements also make it too easy for changes to occur in an uncoordinated fashion during later stages of the project, making it difficult to find the correct solution.

Therefore, the team should develop a requirements document containing, as a minimum, an illustration of the organization's high-level business processes (in other words, how the company or applicable organization(s) operates) and definition of each requirement type. The following list shows the major elements of a requirements document:

- ○ Requirement overview
- ○ Specific requirements
- ○ Constraints
- ○ Assumptions
- ○ Information elicitation methods
- ○ Issues

To produce this document, the team should refer to all information gathered, such as interview and JAD meeting notes, regulations, facility evaluations, and prototypes.

Verifying and Validating Requirements

The importance of requirements can't be overstated—inaccurate requirements lead to solutions that don't adequately support the needs of the users. Thus, the project team should verify and validate the requirements. *Verification* checks if the requirements are accurate based on those needs. *Validation* proves whether the requirements fully represent the needs of the users and conform to the company's business processes. Therefore, these two quality control tests answer these questions:

Verification: "Are we building the product right?"

Validation: "Are we building the right product?"

Verifying Requirements

It's best to verify requirements first, and then validate them. You want to be sure the requirements are okay before testing whether they meet user needs. The most important verification point is to be sure the requirements are complete and unambiguous. Complete requirements describe all aspects of the needs of the users and organization. For example, incomplete requirements might state needs for users and existing systems, but not identify anything about the environment, such as the presence of potential electromagnetic interference. For wireline systems, this might not be critical, but it could have serious impact on the operation of radio-based products. Requirements should be unambiguous to avoid needing clarification later. Ambiguous requirements force the designer to seek the finer details. To save time, most designers will guess the values of the remaining details, causing the designer to choose inappropriate characteristics.

For most projects, you can verify the requirements by referring to the requirements document and answering these questions:

○ Do the requirements address all user and organizational needs?

○ Do the requirements clearly state the needs?

○ Do the requirements avoid describing solutions to the requirements?

Validating Requirements

The best method to validate requirements is to build a prototype as a model that represents the requirements. For application development, you can build a software prototype using a fourth-generation language, such as Powersoft's Powerbuilder, that contains the screens and some functionality to implement the requirements. For off-the-shelf applications and hardware, of course, vendors normally allow enough evaluation time—one or two months—to test the application. For either case, you can have the users exercise the prototype and observe whether their needs will be met.

Baselining Requirements

The baselining, or standardizing, of requirements involves final documenting and approval of the requirements. This process makes the requirements "official," and you should only change them by following an agreed-upon process. Who approves the requirements? Ultimately, the customer representative should give the final sign-off; however, an analyst should endorse the requirements in terms of their accuracy and efficacy. If you're deploying the system under a contract, other people, such as the project manager and contract official, may also need to offer approvals. Be certain to indicate that both the organization and modification team consider the set of requirements as a firm baseline from which to design the network.

Updating the Project Plan

After defining the requirements, it's time to revisit the planning elements you prepared earlier in the project. At first, you probably based the project WBS, schedule, and budget on incomplete and assumed requirements. The actual requirements, though, might cast the project in a different light. For example, maybe you found during the user interviews that information security was more important than you had expected. This might create a need to modify the WBS—and possibly the schedule and budget—to research security technologies and products. Or you might have planned to spend three weeks during installation setting up 150 computers, but during the interviews, you found there will only be 75. This information could enable you to cut back the schedule, or reallocate the time to a task that might take longer than expected.

With an updated project plan and final requirements, the project team is ready to move into the design phase of the project.

CHAPTER 7

Analyzing the Feasibility of a Wireless Network

Implementing a wireless system is generally a costly event. You need to perform a thorough site survey, purchase and install wireless adapters and access points, and possibly procure and install other elements, such as portable computers, hand-held terminals, cabling, and servers. A feasibility study helps organizations decide whether to proceed with the project based on the costs associated with these components and the expected benefits of deploying the system. Before an organization will allocate funding for a project, the executives will want to know what return on investment (ROI) to expect within a particular amount of time. Most companies will not invest a large amount of money, such as $100,000 or more, to deploy a wireless system without the assurance that gains in productivity will pay for the system. Executives should consider the following key factors when making this decision:

○ Costs (including hidden costs)

○ Savings

○ Learning curves

○ Effects on existing systems

Humans are notorious for adapting to change very slowly, or not at all. For instance, there are many benefits in replacing paper-based record systems, such as those used in hospitals and warehouses, with hand-held wireless devices that provide an electronic means of storing and retrieving information from a centralized database. Most people can't make this type of change very quickly. Therefore, executives will need to understand how much time and training the current staff might need before realizing the benefits of the wireless system. Systems managers should be concerned with how the new system will affect the operations and cost of the existing systems.

They will ask questions such as: Will there need to be additional system administrators? Will there be any additional hardware or software maintenance costs?

This chapter addresses the following steps necessary to analyze the feasibility of a wireless network:

○ Performing a preliminary design

○ Building a business case

○ Deciding whether to implement

Performing a Preliminary Design

To figure costs for a project, you need to perform a preliminary design to identify the major system components. The preliminary design provides a high level description of the network, at least enough detail to approximate the cost of implementing and supporting the system. For the preliminary design, you do not need to determine the exact number of connectors and types of network interface cards—you primarily need to decide which technologies to use, such as ethernet versus radio-LAN, and estimate the total cost based on general costs. Later stages of the design phase will further define the components and configurations necessary to implement the system.

> **NOTE**
>
> Refer to Chapter 8, "Designing a Wireless Network," to learn how to accomplish a preliminary design and define the technologies and major components your system will utilize.

Building a Business Case

As shown in figure 7.1, building a business case involves conducting a feasibility study to document the costs and savings of implementing a particular system, as well as offer a recommendation on which direction to proceed. To define costs and savings, you will have to bound the business case; that is, base it on a specific operating time period. Generally, this operating period is the life expectancy of the system. Most organizations are satisfied with a two to three year recovery of benefits on network purchases. Predicting costs and benefits beyond three years can lead to significant margins of error because technologies rapidly change, and most business plans are fairly unstable beyond two years.

```
┌─────────────────────────────────────┐
│          Business Case              │
├─────────────────────────────────────┤
│                                     │
│    • Executive Overview             │
│    • Project Scope                  │
│    • Costs                          │
│    • Savings                        │
│    • ROI                            │
│    • Risks                          │
│                                     │
└─────────────────────────────────────┘
```

Figure 7.1
Elements of a business case.

Executive Overview: Provides a concise overview of the business case.

Project Scope: Defines the resulting wireless system, assumptions, and constraints.

Costs: Details all costs necessary to implement and support the new system.

Savings: Identifies the savings resulting from the deployment and operation of the new system.

Return on Investment (ROI): Describes the difference between the costs and savings of deploying the new system. The ROI is the main factor for basing the decision to proceed with the project.

Risks: Identify the issues that may cause the project to be unsuccessful.

In summary, when developing a business case, you should perform these activities:

○ Recognize applicable feasibility elements

○ Identify costs

○ Identify savings

○ Decide whether to proceed

Recognizing Applicable Feasibility Elements

Figure 7.2 identifies the elements you should consider when building a business case. The goal is to decide which elements apply to the implementation you are undergoing and then to assign costs and savings for each element. Some elements are tangible and some are not. Modification costs, such as the purchase of hardware and software, results in real dollars spent. In addition, increases in productivity result in labor savings or increases in revenue. The computerized image a wireless system brings to a company, however, offers intangible benefits. In the eyes of the customers, for example, a company having a state-of-the-art wireless order-entry system might appear to be superior over other companies with older systems. This does not relate to tangible savings; however, in addition to other factors, it might increase the company's level of business.

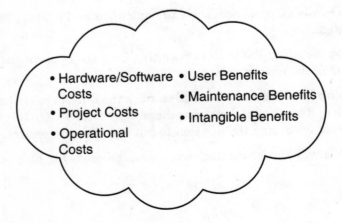

Figure 7.2
Elements to consider when building a business case.

Identifying Costs

When identifying costs, be sure to include everything that the project will require for the implementation and operational support of the system. Do not forget that sustaining the system after it becomes operational will require some funding. Organizations commonly do not include all costs for operational support, such as training and periodic maintenance.

The best format for identifying costs is to utilize a spreadsheet, such as Microsoft Excel, and layout all cost categories and the prices of each. For the cost elements that apply to your project, determine their associated costs, as shown in the following sections.

Hardware and Software Costs

The cost of hardware and software components is one of the largest expenses when implementing a system. These costs include wireless adapters, access points, ethernet boards, network operating systems, application software, cabling, and other components.

Project Costs

Project costs comprise another large percentage of total expenses. Project costs include the labor and materials necessary to complete each phase of the project. These expenses fall into the following categories:

- Planning
- Requirements analysis
- Network design
- Software development
- Operational support preparations
- Installation
- Testing
- Documentation
- Training
- User inactivity

Planning includes costs for scheduling the modification, establishing an implementation team, and periodically revising plans. Software development, if the modification requires it, will consist of the cost of programmers and possibly the purchase of compilers or software development kits. Installation and testing expenses are primarily the cost of technicians and testers, but the team also might need to purchase special tools, such as spectrum analyzers and cable testers, to accomplish their jobs. Documentation is part of every stage of the modification process; therefore, include the price of creating requirement documents, design specifications, schematics, user manuals, and so on. If users are disrupted during the installation of the system, be sure to factor in the cost of their inactivity.

Operational Costs

Once the system is operational, it will cost money to keep it running properly; therefore, include operational expenses over the time period you are basing the business case on. Figure 7.3 illustrates the costs associated with operating the system.

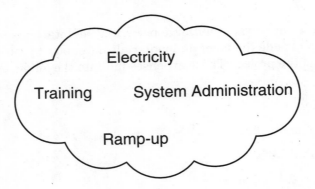

Figure 7.3
Costs associated with operating the system.

○ **Electricity Costs.** The electronic devices within the system, such as computers, network interface cards, servers, and specialized cooling equipment, require electricity; therefore, include a projected cost for the electricity over the applicable time period.

○ **System Administration Costs.** The operational support of the system might require one or more system administrators. These people are needed to maintain user names and passwords, as well as configure printers and back up the files on the server.

○ **Maintenance Costs.** An effective system maintenance organization consists of an adequate set of spare components, documentation, employees, and a facility for the maintenance staff.

○ **Training Costs.** The system might require both initial and recurring training for users and support staff. This results in tuition and possibly travel expenses.

○ **Ramp-up Costs.** In addition to training costs, include other costs associated with migrating to the new system. Initially, user productivity might be low because users normally experience a learning curve when first using the new system. A staff of accountants, for example, might be accustomed to keeping figures on paper and in spreadsheets. A wireless system may utilize a centralized database, allowing the accountants to input and output data directly from a PC. This changes the way that they manage their information, causing a loss in productivity as the they get used to the new system. Over time, employees will become more productive using the database than they were with pencil and paper, but be sure to include the time lost as a cost.

Identifying System Benefits (Savings)

The objective of identifying system benefits is to show how the new system will reduce or avoid costs and increase revenue. Figure 7.4 identifies several areas to focus the analysis of system benefits. Some of these benefits result from lower costs in operating the system, an increase in productivity, faster service, lower maintenance costs, fewer changes to network cabling, improved corporate image, and employee job satisfaction. Other elements deal with the implementation itself, such as less expensive installation in difficult-to-wire areas and reduced installation time.

Figure 7.4
Areas to focus the analysis of system benefits.

Chapter 1, "Introduction to Wireless Networking," describes several benefits of wireless networks, such as mobility, the ability to install in difficult-to-wire areas, reduced installation times, and fewer changes to network cabling. These benefits convert to cost savings when comparing wireless solutions with ethernet or other wireline approaches. Review these benefits in Chapter 1, and use them as a basis for comparison.

The following paragraphs further describe general networking benefits and associated cost savings.

○ **Increased Productivity.** Applications such as file transfer, e-mail, printer sharing, electronic calendaring, networked fax machines, and mobile access to centralized databases and network services enable users to get their tasks done faster, resulting in lower labor costs and higher profits. Increases in productivity equate to lower task completion times, resulting in cost savings based on lower labor hours needed to complete the tasks.

You can easily calculate the cost savings based on an increase in user productivity. Start by determining the amount of time an individual can save by using the new system and multiply this time by the person's pay rate. This equals the cost savings for that individual. An aggregate cost savings can be calculated by adding the savings from all users.

○ **Lower Software Upgrade Costs.** With a network, software upgrades become much faster and less expensive because of the centralized storage of applications. Imagine having 300 stand-alone PCs, and assume someone decides to upgrade an application from one version to another. You could have the users install their own software, but some would not waste their time, others would perform the installation and have trouble, and a few would perform the installation flawlessly. Instead, the best method in this case would be for the system administrator to install the new version of software on all 300 PCs. Assuming an average time of 15 minutes to install the software on each computer, it would probably take this person a couple weeks to install the upgrade. Upgrading software via networked computers is less expensive and less time consuming. In a network, the installer simply installs the new version of software on the server, allowing everyone immediate access to the new upgrade. This only takes 15 minutes, which allows the installer to spend his time working on more important items. To calculate this type of savings, estimate the number of software upgrades that might occur over the applicable period of time and figure the amount of time and dollar savings based on the rate you pay people to install software.

○ **Qualitative Benefits.** Qualitative benefits are based on elements that cannot be assigned specific dollar values. These types of benefits are very important—they often provide an extra incentive to implement a system. A company that develops software, for example, would want to maintain a good corporate image and retain employees by implementing a state-of-the-art network. Otherwise, clients might not consider the company to be a credible software developer. Also, customers of a retail store seeing store staff use wireless terminals to update prices leaves a good impression on the customer.

Documenting the Business Case

Be sure to document all elements of the business case in a form that makes the ROI readily apparent. You can use the elements shown in figure 7.1 as a starting point. Before submitting the business case to the executives for review, assess it according to these criteria:

○ Describes realistic and achievable savings

○ Describes complete and accurate costs

○ Compares costs and savings

○ Clearly explains return on investment

○ Describes issues and risks associated with realizing benefits

○ Is based on a plausible time-frame

○ Provides a recommendation on whether to implement the system or not

Making a Decision to Proceed

The final step is to decide to proceed with the implementation. Distribute the feasibility study to the appropriate managers, and schedule a meeting to discuss the project. Assuming the study convinces management that a strong ROI exists, the decision on how to proceed will be based on the availability of money to fund the project and the presence of implementation issues. Funding constraints and implementation issues, such as weak requirements and solutions, can affect the project schedule.

In some cases, managers might want to divide the project into phases and stagger the implementation over a longer period of time to accommodate the following scenarios:

○ **No funding or implementation issues:** If no funding or implementation issues are present, the company should allocate the funding and press on with a full implementation.

○ **Limited funding and no implementation issues:** If there are no implementation issues and complete funding is not possible, the company could agree to the entire project and spread the deployment out over a time period that accommodates the future availability of money. For example, a company may have 100 sales people located throughout the United States needing mobile access to the company's proposal and contract databases located at the company's headquarters. The proposed wireless system may consist of 100 mobile portable computers, linked to the headquarters' building via CDPD.

Managers may feel strong benefits in providing wireless access to their sales people; however, the existing budget may only be able to fund 50 of the connections (CDPD modems and corresponding service) during the current year. The company may decide to deploy the remaining half of the system the following year.

○ **Implementation issues but no funding issues:** If plenty of money is available but concern exists whether the requirements or design is solid, the company should consider funding only the requirements and design phases of the project to better clarify the needs and the solution. This will increase the accuracy of the cost estimate associated with hardware, software, and support. It also ensures the purchase of the right components. For example, the business case may do a good job of identifying the benefits and savings a company will receive by deploying the system, but it may not have been possible to define a solution that would provide assurance of component costs or whether a solution even exists. In this case, the company should fund enough of the project to accurately define components necessary to satisfy the requirements. This would enable the company to make a better decision later to allocate money for component procurement and the installation phase of the project.

○ **Limited funding and implementation issues:** If funding is limited and there are issues with implementing the system, the company should not continue with the project or proceed with extreme caution. For example, there may be fantastic benefits in deploying a wireless patient record system in a hospital, but limited funding and the presence of implementation issues, such as potential interference with medical instruments and the task of migrating existing paper-based records into a database, should cause the organization to think twice before funding the project. In this case though, the company could fund smaller projects to resolve the issues and reconsider the implementation of the system at a later date.

PART III

Implementing and Supporting Wireless Networks

8. Designing a Wireless Network

9. Planning the Support of a Wireless Network

10. Installing a Wireless Network

Designing a Wireless Network

After fully defining the requirements for a system, the next step is to perform the design, which determines the technologies, products, and configurations providing a solution. As with any engineering activity, the goal of network design is to find a solution that meets requirements at the least cost. In some cases, a preliminary design can establish a basis for initial cost estimating and work planning. But the design phase of the project defines all aspects of the solution and supports product procurement, installation, testing, and operational support of the system.

The design phase of the project produces items such as schematics, building layout drawings, bills of materials (parts lists), and configuration drawings. These items are necessary to fully define the design.

This chapter discusses the following steps, which are necessary for completing the design of most projects:

- ○ Define network elements
- ○ Identify products
- ○ Identify the location of access points
- ○ Verify the design
- ○ Document the final design
- ○ Procure components

Defining Network Elements

The process of defining network elements includes deciding which technologies and standards to utilize as a solution to the requirements. You may decide, for example, to use an ethernet network to provide connectivity between access points. In some cases, you might also need to select a product to define the network element. This is mostly applicable with the network operating system and off-the-shelf applications. For example, you might choose Novell Netware as the network operating system. For word processing, you might decide to use Microsoft Word.

The general process of defining network elements is as follows:

1. Identify which network elements apply

2. Determine values for each network element

In other words, identify that part of the network architecture on which you need to concentrate to find a solution, and then determine the technologies, standards, and, if necessary, the products for each element.

Identifying which Network Elements Apply

A network consists of many elements that support the dissemination of information among applications, databases, and systems. Figure 8.1 identifies the elements that comprise a network, from application to the physical layer. The goal of this step is to identify the network elements that apply to your specific implementation. To accomplish this, review the network requirements and develop the list of elements that you will need to consider.

Determining Values for Network Elements

The next step in defining the network elements is to determine values for each element. As with most engineering efforts, it is a good idea to maximize the use of standards when selecting technologies and to use a top-down approach.

Ideally, the design team should select technologies and standards having the highest level of maturity. Careful selection leads to longer lasting and easier-to-maintain solutions. When assessing maturity levels, use the following evaluation criteria:

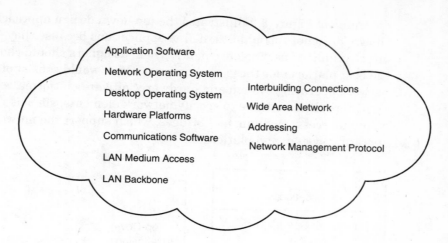

Application Software

Network Operating System

Desktop Operating System

Hardware Platforms

Communications Software

LAN Medium Access

LAN Backbone

Interbuilding Connections

Wide Area Network

Addressing

Network Management Protocol

Figure 8.1
The identification of network elements.

Low Maturity

○ No standard and low product proliferation

○ De facto standard and low product proliferation

○ Emerging official standard and low product proliferation

○ Stable official standard, but reaching obsolescence; low or high product proliferation, with vendors and end-users switching to other technologies

Moderate Maturity

○ No standard and high product proliferation

○ Emerging official standard and high product proliferation

High Maturity

○ Stable official standard and high product proliferation

○ De facto standard and high product proliferation

A top-down design approach first defines high-level specifications directly satisfying network requirements, then defines the remaining elements in an order that mostly satisfies specifications already determined. Figure 8.2 illustrates the top-down design approach. As shown, designers should first define applications because they directly support user requirements. Then, designers should choose the best platform for these applications, which would consist of the network operating system and hardware/software platforms. Designers then continue to specify network elements, such as medium access, medium, and bridges, which support the network operating system and platforms.

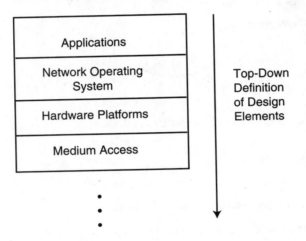

Figure 8.2
The top-down design approach.

The following sections offer potential technologies, standards, and products for each of the network elements. Review these sections and, using a top-down approach, choose the solution that best meets your requirements.

Application Software

Users can accomplish required tasks, such as word processing, database access, and e-mail, with application software. Application software, then, directly satisfies network requirements, particularly user requirements. There are many types of applications, ranging from simple utilities to fully-featured office automation. The following briefly describes various categories of applications:

○ **Vertical Market Applications.** Provide customized data entry, query, and report functions for various industries, such as healthcare and banking.

○ **Office Software.** Generally consists of word processing, database, spreadsheet, graphics, and electronic mail.

○ **Scientific Applications.** Provide the analysis of real-world events by simulating them with mathematics.

○ **Project Management Software.** Renders the capability to efficiently track a project and analyze the impact of changes.

○ **CAD (Computer Aided Design) Software.** Uses vector graphics to create complex drawings.

○ **Desktop Publishing Software.** Provides the capability to effectively merge text and graphics and maintain precise control of the layout of each document page.

○ **Infoware.** Provides online encyclopedias, magazines, and other references.

○ **Mathematical Programs.** Enable the creation and execution of complex mathematical equations.

○ **Multimedia Software.** Adds graphics, sound, and video for use in education and specialized applications.

Most traditional applications are hosted centrally on mainframe computers, and users access these applications via dumb terminals or PCs running terminal emulation software. In this case, the application runs entirely on the mainframe computer. The user interfaces tend to be character-oriented instead of graphical, making them somewhat user unfriendly. Another problem is that they are costly and time-consuming to change. For example, a programmer normally must alter the program to accommodate simple changes to a report format. Some products, though, are starting to appear on the market that convert terminal updates to graphical formats. For example, Client/Server Technologies Inc.'s GUISys/3270 file format conversion software enables users to migrate from IBM's 3270 screens to GUI applications. GUISys/3270 is a Microsoft Windows application that reads 3270 screen information from a PC's 3270 emulator buffer and sends the data through a pattern-matching knowledge base to create GUI formats for the

3270 screen. This product enables users to change screens without complex programming techniques. All-in-all, however, mainframes are generally more expensive to maintain.

Today, many companies are beginning to deploy client-server applications based on the model shown in figure 8.3. The user interfaces to the application located on the server via a client program that performs some of the processing. The server software is generally a database, such as Sybase.

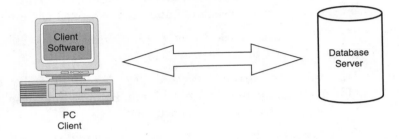

Figure 8.3

The concept of a client-server system.

There are several benefits of developing applications based on client-server principles. For one, the interfaces are graphical (generally based on Microsoft Windows) and are very simple to use and relatively easy to develop and modify. For example, Powersoft's Powerbuilder and Microsoft's Visual Basic are common tools for developing client-server applications that interface with various databases.

Many applications are available off-the-shelf, especially common office software such as Microsoft's Word for word processing, PowerPoint for presentations, and Excel for spreadsheets. In some cases, though, it might be necessary to develop a specific application. This process is beyond the scope of this book, but would include a software engineering process similar to deploying a wireless network—that is, to analyze requirements, design the software, implement the code, and install. The main difference between an application and network implementation project is the level of detail required in the analysis phase. Generally, you must determine much more specific requirements for applications than for the networks that support them.

Network Operating System

The network operating system (NOS) provides a platform for applications, printing, and file sharing. The NOS, combined with a hardware platform such as a PC, is referred to as a server. Most NOSs require the installation of client software on a user's PC to act as an interface to the server. As shown in figure 8.4, some NOSs are server-oriented, meaning a dedicated hardware platform supports the NOS software and users generally access network services directly from the server. Generally the server-oriented NOSs offer the greatest performance and functionality. Other NOSs are peer-to-peer, which distribute network applications and services among the users' workstations. Novell's NetwareLite and Artisoft's Lantastic are examples of peer-to-peer NOSs.

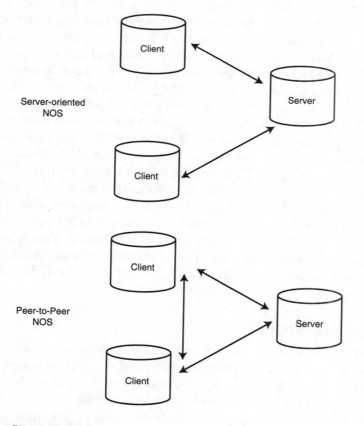

Figure 8.4

Server-oriented versus peer-to-peer network operating systems.

Through the NOS software, a system administrator can open accounts for users by assigning user names, passwords, and access rights to resources located on the server. The NOS supports the remote storage of applications and files, and users can access these via the network. If a user runs an application, such as Microsoft Word, located on a server, the user opens the drive mapped to a directory on the server where the application resides, then runs the application. The server sends the application to the user's workstation where the application will run. As users create documents, they can store the files on the server.

In some cases, such as the interfacing of wireless terminals to a database, you probably do not need an NOS—all the networking you need can be handled by wireless products performing physical and data link functions, along with software to interface with the database. If you're going to host client-server applications for others to access, you need to choose an NOS. Two prominent NOSs on the market today are Microsoft's NT Server 3.51 and Novell Netware 4.1. Novell's Netware supports access to all enterprise network resources via a single-point login and leads the industry in market share for NOSs. Most people agree Netware is an excellent choice for file and print services; however, many feel Netware doesn't support applications as well as NT Server. NT Server is a 32-bit multithreaded, multitasking operating system and is scaleable in terms of processors. NT Server also includes a preemptive multitasking operating system that dedicates time slots to application processes. This means that the failure of one process does not affect the processing time of another process; thus NT Server is unlikely to crash because of a single bad application. This is analogous to the operation of a multicylinder automobile engine. If one of the spark plug wires breaks, the other spark plugs keep the car running.

Desktop Operating System

Whether the wireless network implementation includes either portable or desktop PCs, you probably need to select an appropriate desktop operating system. This operating system runs on the PC and supports the client piece of the NOS and client-server application. Desktop operating systems have undergone a dramatic evolution in the past fifteen years. Initially for PCs, Microsoft developed the Disk Operating System (DOS), which is character-based and requires users to understand DOS commands in order to copy and edit files and run applications. Later, of course, Microsoft introduced Windows, offering a graphical interface for users to manage their files and execute applications.

Today, the main choices for PC operating systems are Microsoft Windows 3.1, Windows for Workgroups, Windows NT, and Windows 95. The following briefly describes each of these operating systems:

○ **Microsoft's Windows 3.1.** In 1992, Microsoft introduced Windows 3.1, which provided a more stable environment for running 16-bit Windows and DOS applications than its predecessor, Windows 3.0. Windows 3.1 supports multimedia, TrueType fonts, compound documents (OLE), and drag-and-drop capabilities. Windows 3.1 also runs 32-bit Win32s applications by translating them into 16-bit operating system calls.

○ **Microsoft's Windows for Workgroups.** Microsoft's Windows for Workgroups (WFW) is a version of Windows 3.1 that includes built-in peer-to-peer networking and electronic mail. WFW also includes 32-bit file access, which bypasses DOS and replaces SmartDrive with another disk cache for increased performance. WFW includes integrated file sharing, electronic mail (Microsoft Mail) and workgroup scheduling (Schedule+).

○ **Microsoft's Windows NT.** Microsoft's Windows NT (New Technology), introduced in 1993, is an advanced 32-bit operating system for Intel 386s and up, MIPS (described later in "Hardware Platforms"), Alpha, and PowerPC CPUs. NT is a self-contained operating system that does not use DOS, and it runs NT-specific applications as well as DOS and Windows applications. Windows NT features include peer-to-peer networking, preemptive multitasking, multithreading, multiprocessing, fault tolerance, and extensive security. Windows NT supports 2 GB of virtual memory for applications and 2 GB for its own use.

○ **Microsoft's Windows 95.** Windows 95 offers a user interface with folder metaphor, explorer view, and taskbar, providing end users with a more intuitive and easy-to-use Windows platform. The Windows Setup program auto-senses PC peripherals. Windows 95 supports all popular LAN protocols, including TCP/IP, IPX/SPX, and NetBEUI. It also provides Remote

LAN connectivity through Microsoft's Remote Access Server (RAS). Internet access and on-line access are available as add-ons. Windows 95 supports the majority of Windows 3.x applications and existing Intel x86 hardware, as well as 32-bit applications.

Hardware Platforms

A computer hardware platform consists of a central processing unit (CPU), memory, applicable interfaces, and an operating system. If you are deploying an NOS or user workstations, you need to specify a hardware platform for each. Typical hardware platforms include the 386, 486, and Pentium PCs.

As a performance measure, vendors classify their platforms according to the rate at which the machine can process instructions. This unit of measurement is MIPS (Million Instructions Per Second). High-speed personal computers, such as Pentiums, are usually capable of operating at 100 MIPS or greater. A 386 PC usually runs between 3 to 5 MIPS. However, MIPS rates are not uniform because some vendors use the best-case value of the platform and others use average rates. As a result, designers should not consider MIPS as the single factor when sizing up the performance of a hardware platform. Be sure to consider other attributes as well, such as bus and memory speed, memory management techniques, and the operating system.

As you have probably encountered before, there are a large number of options available for personal computers. How do you choose the best platform for your requirements? The best way is to start with the minimum requirements specified by the application and NOS vendors. In most cases, vendors test and certify the operation of their software on various PC configurations. It is a good idea to ask the vendor for recommendations on which platform to purchase.

Communications Software

Once you select applications and a server, you need to specify communications software that interfaces users with the applications. For applications located on the local server, the NOS generally provides this functionality for local users. However, users needing to access resources located at different sites via the

Internet or located on a mainframe computer will probably require Transmission Control Protocol (TCP) and/or 3270 emulation software.

Transmission Control Protocol

TCP is a commonly used protocol for establishing and maintaining communications between applications on different computers, and provides full-duplex, acknowledged, and flow-controlled service to upper-layer protocols and applications. For example, a mainframe computer employing TCP software enables a wireless user, having TCP software as well, to log in to the mainframe using Telnet and run the application.

LAN Medium Access

As part of the network design, you need to specify a particular network interface card that provides access to a medium that interconnects the computers.

As described in Chapter 2, "Wireless Local Area Networks (LANs)," there are two common types of medium access: carrier sense access and token passing. Most wireless LAN products today deploy a proprietary form of carrier sense access. Vendors, however, will probably migrate their products to the IEEE 802.11 medium access technique, which is based on a carrier sense access process, once that standard becomes finalized in 1997. Some infrared products use a directed beam to create a token-ring network. With the implementation of a wireless LAN, you need to choose one of these methods.

The choice of medium access depends mainly on information flow requirements. Carrier sense protocols will satisfy most requirements; however, they offer asynchronous, statistical access. In other words, you cannot predict when stations will be able to send data. Therefore, carrier sense networks don't support the transmission of real-time data, such as voice and video, very well. On the other hand, token passing networks operate in a synchronous manner, offering better support for synchronous, real-time data transfers.

In addition to the wireless LAN technologies and products discussed in Chapter 2, you might need to consider several wire-based standards as part of your implementation process. For example, it may be beneficial to locate a database server on a network with both wired connections to stationary users and wireless access points. As a result, you will need to decide which wired access method is best to use as the backbone network. The three main wire-based LAN standards are:

○ IEEE 802.3 (CSMA)

○ IEEE 802.5 (token ring)

○ ANSI Fiber Distributed Data Interface (FDDI)

The IEEE 802.3, 802.5, and 802.11 specifications are part of the overall IEEE standards hierarchy (see fig. 8.5). IEEE defines a LAN with three layers of functionality: Logical Link Control (LLC), Medium Access Control (MAC), and physical layers. The LLC, which is IEEE 802.2, provides link synchronization, and the MAC layer is responsible for medium access. The physical layer defines electrical characteristics of the signal and medium. Thus, 802.3, 802.5, and 802.11 not only specify the medium access method but also the type of medium. The LAN Backbone section in this chapter provides more details on possible physical layers offered by 802.3 and 802.5.

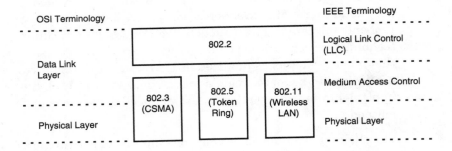

Figure 8.5
The IEEE 802 standards hierarchy.

IEEE 802.3 Carrier Sense Multiple Access

The IEEE 802.3, released in 1980, is the most popular access method. It operates at 10 or 100 Mbps, depending on the type of physical layer used. The use of this standard satisfies most performance requirements, and, because of its high degree of proliferation, products are low cost compared to token ring and FDDI.

Therefore, consider IEEE 802.3 as your wired network medium access standard, unless you require real-time data transfers or a high bandwidth. For those needs, you may want to consider IEEE 802.5 or FDDI.

IEEE 802.3 is based on the ethernet protocol developed by Xerox Corporation's Palo Alto Research Center (PARC) in the 1970s. Shortly after the release of the 802.3 standard, Digital Equipment Corporation, Intel Corporation, and Xerox Corporation jointly developed and released a very similar ethernet specification (Version 2.0). Today, most organizations use the IEEE 802.3 specification, which is commonly called ethernet, as a basis for their LANs. As shown in figure 8.6, IEEE 802.3 and the ethernet specification describe a slightly different frame header; therefore, these protocols are not compatible.

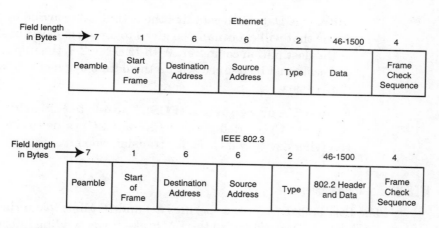

Figure 8.6
The frame headers of IEEE 802.3 and ethernet.

The following describe each of the 802.3 and ethernet frame header fields:

○ **Preamble.** Both ethernet and IEEE 802.3 frames begin with an alternating pattern of ones and zeros called a preamble, which tells receiving stations that a frame is coming.

○ **Start-of-Frame (SOF).** The SOF delimiter ends with two consecutive one bits, which serve to synchronize the frame reception portions of all stations on the LAN.

○ **Destination and Source Address.** The Destination and Source Address fields are 6 bytes long and refer to the addresses contained on the ethernet and IEEE 802.3 network interface cards. The first 3 bytes of the addresses are specified by the IEEE on a vendor-dependent basis, and the last 3 bytes are specified by the ethernet or IEEE 802.3 vendor. The Source Address is always a unicast (single node) address. The Destination Address may be unicast, multicast (group), or broadcast (all nodes).

○ **Type.** Ethernet frames have a 2-byte Type field that specifies which upper-layer protocol will receive the data after ethernet processing is complete.

○ **Length.** The Length field in IEEE 802.3 frames indicates the number of bytes of data in the Data field.

○ **Data.** The Data field contains the actual data carried by the frame that will eventually be given to an upper-layer protocol at the destination computer. With IEEE 802.3, the upper-layer protocol must be defined within the data portion of the frame if necessary.

○ **Frame Check Sequence (FCS).** The 4-byte FCS field contains a Cyclic Redundancy Check (CRC) value so the receiving device can check for transmission errors.

IEEE 802.5 Token Ring

The IEEE 802.5 standard specifies a 4 and 16 Mbps token ring LAN. Stations connected to the LAN take turns sending information to other stations by utilizing a token as explained in Chapter 2 within the section "Point-To-Point Infrared LAN System." Because of the token access method, 802.5 supports heavier traffic under more stable conditions than 802.3 ethernet. In addition, 802.5 supports deterministic access to the medium, which enables it to handle synchronous type information transfers. IEEE 802.5 is the second most popular LAN medium access technique and is more expensive to implement than ethernet. But, as mentioned earlier, use token ring in cases where you need better performance.

The first token-ring network was developed by IBM in the 1970s; then IEEE wrote the IEEE 802.5 specification based on IBM's work. Today, IBM Token Ring and IEEE 802.5 networks are compatible, although there are minor differences. For instance, IBM's Token Ring network specifies a star configuration with all end stations attached to a device called a multistation access unit (MSAU). IEEE 802.5 does not specify a topology, but most 802.5 implementations are also based on a star configuration. Also, IEEE 802.5 does not specify a media type, but IBM Token Ring identifies the use of twisted-pair wire.

The following section explains the purpose of each field.

Tokens

○ **Start Delimiter.** The Start Delimiter alerts each station that a token (or data frame) is arriving. This field includes signals that distinguish the byte from the rest of the frame by violating the encoding scheme used elsewhere in the frame.

○ **Access Control.** The Access Control byte contains the priority and reservation fields, as well as a token bit that is used to differentiate a token from a data or command frame. The monitor bit is used by the active monitor to determine whether a frame is endlessly circling the ring.

○ **End Delimiter.** The End Delimiter identifies the end of the token or data/command frame. It also contains bits to indicate a damaged frame, as well as the last frame in a logical sequence.

Data/Command Frames

○ **Frame Control.** The Frame Control byte indicates whether the frame contains data or control information. In control frames, the Frame Control byte specifies the type of control information.

○ **Destination and Source Address.** As with IEEE 802.3, the Destination and Source Address is 6 bytes long and designates the source and destination stations.

○ **Data.** The Data field contains the data being sent from source to destination. The length of this field is limited by the token holding time, which defines the maximum time a station may hold the token.

○ **Frame Check Sequence (FCS).** The 4-byte FCS field contains a Cyclic Redundancy Check (CRC) value so the receiving device can check for transmission errors.

○ **End Delimiter.** The End Delimiter identifies the end of the data/command frame. It also contains bits to indicate a damaged frame, as well as the last frame in a logical sequence.

ANSI Fiber Distributed Data Interface

The Fiber Distributed Data Interface (FDDI) standard was produced by the ANSI X3T9.5 standards committee in the mid-1980s, and it specifies a 100 Mbps dual token-ring LAN. FDDI specifies the use of optical fiber medium and will support simultaneous transmission of both synchronous and prioritized asynchronous traffic. The CDDI (Copper Data Distributed Interface) version of FDDI operates over Category 5 twisted-pair wiring. FDDI is an effective solution as a reliable high speed interface within a LAN or corporate network.

FDDI is an expensive, but effective solution for supporting high speed deterministic access to network resources. Some organizations find it necessary to use FDDI to connect a group of servers in a server pool. It is also beneficial to use FDDI as the backbone for a campus or enterprise network. The synchronous mode of FDDI is used for those applications whose bandwidth and response time limits are predictable in advance, permitting them to be pre-allocated by the FDDI Station Management Protocol. The asynchronous mode is used for those applications whose bandwidth requirements are less predictable or whose response time requirements are less critical. Asynchronous bandwidth is instantaneously allocated from a pool of remaining ring bandwidth that is unallocated, unused, or both.

ANSI is currently developing FDDI II, which is an extension of FDDI. FDDI II will have two modes: The Basic Mode (which is FDDI) and the Hybrid Mode that will incorporate the functionality of basic mode plus circuit switching. The addition of circuit switching will enable the support of isochronous traffic. Isochronous transmission is similar to synchronous, but with isochronous, a node is capable of sending data at specific times. Decreased source buffering and signal processing simplify the transmission of real-time information.

LAN Backbone

After selecting the medium access method(s), you need to decide the components of the LAN backbone, which include medium, hubs, switches, and access points. Part of this effort will also consist of showing how the components interface with each other. Therefore, you also need to specify the topology and configuration of the LAN.

The medium type fulfills the functionality of the physical layer of either the OSI or IEEE reference models. As described in Chapter 2, medium choices for wireless LANs are radio waves and infrared light. Recall that radio waves are capable of penetrating walls and can provide ranges of up to 1,000 feet, depending on the construction of the facility. However, radio waves are susceptible to interference from other systems. Infrared light will not penetrate walls, limiting the range to a single room or factory floor. Also, infrared light-based LANs do not support mobility (constant movement) very well. Most other systems, however, will not interfere with infrared light signals. Like all design elements, the choice of wireless medium depends on requirements. Radio waves will satisfy most requirements; however, there could be some unique requirements that point you toward the use of infrared light.

Choosing Wireless Media

Here are some suggestions on how to choose a wireless media:

Radio Waves

Consider the use of radio waves for the following situations:

- Users requiring mobility; that is, they need to move while accessing network resources.

- Users requiring access to network resources throughout the building.

Infrared Light

Consider the use of infrared light for the following situations:

- Radio wave interference is a potential problem.

- It is desirable to contain the communications signals within a closed area.

- Requirements identify the need to support high-speed, synchronous data transfers. (In this case use direct, token passing infrared.)

Choosing Physical Media

In addition to the wireless media, you might need to specify the use of physical media as well, especially if the network will include IEEE 802.3, IEEE 802.5, or FDDI wireline technologies. Here is an overview of these media types:

○ **Twisted-pair Wire.** Twisted-pair wire uses metallic conductors, providing a path for current flow. The wire is twisted in pairs to minimize the electromagnetic interference resulting from adjacent wire pairs and external noise sources. A greater number of twists per foot increases noise immunity. Twisted-pair wiring is inexpensive to purchase and easy to install, and it is currently the industry standard for wiring LANs. IEEE 802.3, IEEE 802.5, and ANSI FDDI specify the use of unshielded twisted-pair (UTP) wiring. Consider the use of twisted-pair for wired connections inside buildings, unless you need a higher degree of noise immunity or information security.

The EIA 568 building wiring standard specifies the following five categories of unshielded twisted-pair wiring:

○ **Category 1.** Old-style phone wire, which is not suitable for most data transmission. This includes most telephone wire installed before 1983, in addition to most current residential telephone wiring.

○ **Category 2.** Certified for data rates up to 4 Mbps, which facilitates IEEE 802.5 Token-Ring networks (4 Mbps version).

○ **Category 3.** Certified for data rates up to 10 Mbps, which facilitates IEEE 802.3 10baseT (ethernet) networks.

○ **Category 4.** Certified for data rates up to 16 Mbps, which facilitates IEEE 802.5 Token-Ring networks (16 Mbps version).

○ **Category 5.** Certified for data rates up to 100 Mbps, which facilitates ANSI FDDI Token-Ring networks.

The 10 Mbps version of IEEE 802.3, for example, requires at least Category 3 cable or higher, while the copper-based version of FDDI requires Category 5 cable. There is very little difference in price between Category 3 and 5 cable, and labor costs to install each are the same. Therefore, you should install Category 5 cable for all wired-network installations, regardless of whether you need the extra bandwidth or not. This will avoid expensive re-wiring if you require higher performance in the future.

IEEE 802.3 describes several physical layer specifications, namely 10baseT and 100baseT. The word "base" identifies the use of baseband signals—that is, digital signals. The "10" and "100" mean 10 Mbps and 100 Mbps, respectively. 10baseT and 100baseT configure network devices as shown in figure 8.7. Both enable computers to connect in a star topology, via IEEE 802.3 network interface cards (ethernet boards), to a hub or switch at a distance of 100 meters (300 feet). A hub normally deploys "shared ethernet" that constitutes a single-collision domain. As an example, when station A transmits a frame, all other stations connected to that hub (B and C) will receive it. Since ethernet operates in a half-duplex mode, station A will block all other stations connected to the hub (within the collision domain) from transmitting. A switch, though, is smarter because it physically connects the sending station directly to the receiving station only. This results in multiple collision domains that significantly increase throughput. For example, communications can take place simultaneously between stations A and H and stations D and G.

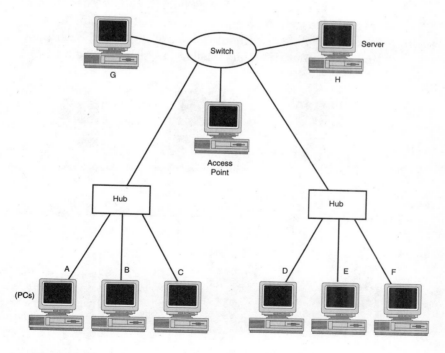

Figure 8.7
The configuration of 10baseT and 100baseT networks.

○ **Coaxial Cable.** Coaxial cable includes a solid metallic core with a shielding as a return path. The shielding reduces electrical noise interference within the core wire. As a result, coaxial cable can extend to much greater lengths than twisted-pair wiring. The disadvantage of coaxial cable is its bulky shape, which makes it difficult to install. Also, coaxial cable does not lend itself very well to centralized wiring topologies, making it difficult to maintain.

During the 1980s, coaxial cable was very popular for wiring LANs, and you might find some still existing in older implementations. Very few—if any—new implementations will require the use of 10base2 or 10base5; however, you should be aware of these types of networks in case you have wireless users that need to interface with them. IEEE 802.3 defines two physical layer specifications, 10base2 and 10base5, based on the use of coaxial cable. 10base2 uses RG58 cable, the same used to connect your television to a cable outlet, and will operate at a distance of up to 185 meters (600 feet). 10base5 uses a much larger cable than RG58, but is capable of operating up to 500 meters (1,640 feet) without the use of repeaters. Both 10base2 and 10base5 utilize a bus topology, as shown in figure 8.8.

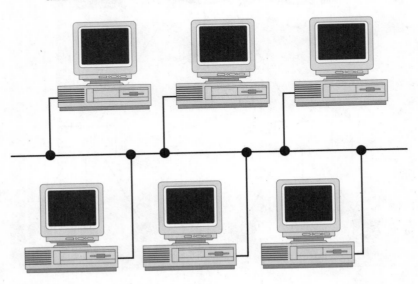

Figure 8.8
The configurations of 10base2 and 10base5 networks.

○ **Optical Fiber.** Optical fiber is a medium that uses changes in light intensity to carry information from one point to another (see fig. 8.9). An optical fiber system consists of a light source, optical fiber, and light detector. A light source changes digital electrical signals into light (on for a logic "1" and off for a logic "0"), the optical fiber transports the light to the destination, and a light detector transforms the light into an electrical signal.

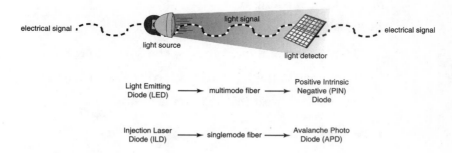

Figure 8.9
The optical fiber communication system.

The main advantages of optical fiber are very high bandwidth (Mbps and Gbps), information security, immunity to electromagnetic interference, lightweight construction, and long distance operation without signal regeneration. As a result, optical fiber is superior for bandwidth demanding applications and protocols, operation in classified areas and between buildings, as well as installation in airplanes and ships. IEEE 802.3's 10baseF and 100baseF specifications identify the use of optical fiber as the medium. FDDI identifies the use of optical fiber as well.

Interbuilding Connections

The interconnection of buildings often goes beyond the range of wireless LAN technologies and standards. Therefore, you need to use other means to provide these types of connections. In general, the wireless technologies offer many advantages as noted in Chapter 1, "Introduction to Wireless Networking," for installation of links between buildings in difficult-to-wire environments, such as across freeways, rivers, or hard soil. For wireless implementations, consider the use of radio-based or infrared-based approaches as described in Chapter 3, "Wireless Metropolitan Area Networks (MANs)." Recall that these methods use highly directive signals to focus the power in a single direction.

Choosing Interbuilding Media

The following are some suggestions for choosing a wireless interbuilding medium type:

Radio Waves

Consider the use of radio waves for the following situations:

○ Buildings that are separated farther than 1 mile

○ When the lowest cost solution is desired

Infrared Light

Consider the use of infrared light for the following situations:

○ When high bandwidth is required

○ When the potential for radio frequency interference is high

○ When buildings are separated by less than one mile

○ When information security is important

Optical Fiber

Optical fiber is an alternative to wireless networking. In many cases, your local city, county, or building company might have installed multiple optical fiber strands between buildings in your area. In that case, building-to-building interconnectivity would only require you to obtain access to the fiber in both of the buildings. But if no fiber exists between the buildings of interest, a wireless solution will probably be most practical.

Wide Area Networks

If requirements indicate the need for mobile users covering a large geographical area to have access to network resources, then you certainly want to focus on one of the wireless WAN approaches mentioned in Chapter 4, "Wireless Wide Area Networking (WANs)."

These approaches—packet radio, analog cellular, CDPD, satellite, or meteor burst—are all based on the use of radio waves. Refer to Chapter 4 for trade-offs between these wireless methods when selecting a wireless WAN technology. In many cases, though, you may be only deploying wireless LAN components. Requirements stating the need for high bandwidth and operation from fixed sites will necessitate the design and implementation of a wire-based WAN. The remaining part of this section briefly describes traditional WAN concepts and technologies.

A WAN consists of routers and links. As previously mentioned, routers receive a routable data packet, such as Internet Protocol (IP) or Internetwork Packet Exchange (IPX), review the destination address located in the packet header, and decide which direction to send the packet next in order to forward it closer to the final destination. The routers maintain routing tables that adapt, via a routing protocol, to changes in the network. Refer to a later section ("Routing Protocols") for a description of common routing protocols.

WANs fall into two main classes: private and public. Figure 8.10 illustrates these two approaches.

Private Point-to-Point WAN

With private WANs, the user organization or company owns and manages most of the network equipment, such as routers and communications circuits. Traditionally, organizations have implemented private point-to-point WANs to support communications between remote terminals and centralized mainframe-based applications. The following identifies the ramifications of using a private WAN:

○ More suitable for a WAN requiring a low degree of meshing (centralized topology)

○ Economically feasible service fees for metropolitan areas

○ Higher initial cost for a greater number of hardware interfaces and circuit installations

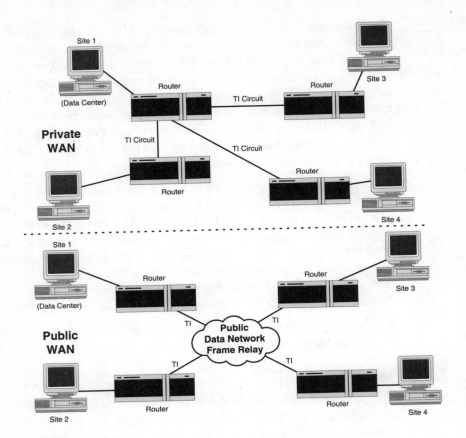

Figure 8.10
Private versus public WAN approaches.

○ Lease fees sensitive to the distance between sites (costs increase as the distance increases)

○ Potentially higher operating costs due to required in-house management

○ Fixed bandwidth

A common link between sites of a private point-to-point LAN is T1. T1 Bell labs originally developed T1 to multiplex multiple phone calls into a composite signal, suitable for transmission through a digital communications circuit. A T1 signal consists of a serial transmission of T1 frames, as shown in figure 8.11. Each frame transports an 8-bit sample of 24 separate channels. You can lease from a telephone service carrier an entire T1 circuit (1.544 Mbps) or only single channels (64 Mbps each).

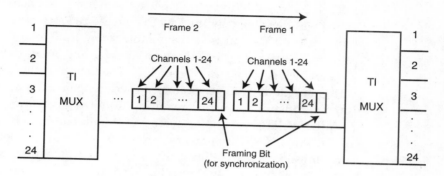

Figure 8.11
The T1 signal format.

Public Packet Switching WAN

A public WAN is owned and operated entirely by a service provider. With the development of distributed client-server applications, most organizations now require technologies suitable for highly meshed topologies. In other words, there is a need to support communications among the remote sites, not just in a centralized data center. Thus, you should seriously consider leasing the use of a public packet switching WAN to support today's demand for distributed computing. The following identifies the implications of using a public WAN:

○ More suitable for a WAN requiring a high degree of meshing (i.e., distributed topology)

○ Lower initial cost

○ Potentially lower operating cost

○ Lease fees not sensitive to the distance between sites

○ Variable bandwidth (bandwidth on demand)

○ Potentially lower operating costs due to carrier-provided management

○ Most economical fees for service outside the metropolitan area

The following briefly describes each of the technologies that support public packet switching WANs. You can lease these as services from carriers such as AT&T and Sprint within most metropolitan areas.

○ **X.25.** X.25 was the first public packet switching technology, which was developed by the CCITT and offered as a service during the 1970s that is still available today. X.25 offers connection-oriented (virtual circuit) service and operates at 64 Kbps, which is too slow for some high-speed applications. Designers made X.25 very robust to accommodate the potential for transmission errors resulting from transport over the metallic cabling and analog systems used predominately in the 1970s. Thus, X.25 implements very good error control. Some companies have a significant investment in X.25 equipment and are still supporting the technology. However, you should consider other packet switching technologies, such as frame relay, for new implementations.

○ **Frame Relay.** Frame relay is a packet switching interface that operates at data rates of 56 Kbps to 2 Mbps. Actually, frame relay is similar to X.25, minus the transmission error control overhead. Thus, frame relay assumes a higher layer, end-to-end, protocol to check for transmission errors. Carriers offer frame relay as a permanent connection-oriented (virtual circuit) service. In the future, frame relay will be available as a switched virtual circuit service. To interface with frame relay service, you need to purchase or lease a Frame Relay Attachment Device (FRAD) or router with a frame relay interface. The FRAD or router interfaces a LAN (typically ethernet) to the local frame relay service provider via a T1 circuit. Frame relay is currently the most feasible technology available for interconnecting geographically disparate sites, especially if these sites span several metropolitan areas and applications are distributed.

○ **Switched Multimegabit Data Service (SMDS).** Switched Multimegabit Data Service (SMDS) is a packet switching interface that operates at data rates ranging from 1.5 Mbps to 45 Mbps. SMDS is similar to frame relay, except SMDS provides connectionless (datagram) service. You can access a local SMDS service provider via T1 or T3 (45 Mbps) circuits. SMDS is not available in all areas.

○ **Integrated Services Digital Network (ISDN).** During the 1980s, the ISO developed a set of standards for the Integrated Services Digital Network (ISDN). The goal of the ISDN is to offer worldwide multimedia services via a single standard network connection, which means having one connection to television, radio, telephone, and computer networks. This goal has never been met, but many carriers offer data communications services via ISDN interfaces. Today, you can lease an ISDN circuit that operates at 64 Kbps using digital signals and optical fiber circuits. This supports telephone and data traffic requiring synchronous transmission, such as video conferencing.

Information Security

As explained in Chapter 1, "Introduction to Wireless Networking," wireless networks are prone to breaches of information security. When sending data over a wireless network, leased lines, or a public data network, you run the risk of someone stealing the information. If you plan to send sensitive information over a WAN, consider the use of encryption and other mechanisms to counteract this problem. For example, you can use encryption devices on all links to scramble the data, making it meaningless. For connections to the Internet, place a firewall at the entry point to the Internet. A firewall filters out unauthorized access from others on the Internet to your company's resources. A typical firewall configuration blocks all incoming traffic, but enables users from within to access Internet services.

Addressing

If you plan to use applications requiring TCP/IP interfaces or if users need access to the Internet, you have to assign a unique Internet Protocol (IP) address to each device connected to the network (workstation, printer, server, and so forth). Actually, the IP address corresponds to a network connection, so a server having two network interface cards would require two IP addresses—one for each card.

The IP packet header contains the source and destination IP address that routers will use, along with a routing table, to determine where to send the packet next. The IP address is 32 bits long, allowing 4,294,967,296 unique IP addresses.

The developers of the Internet based IP addressing on the hierarchical format shown in figure 8.12, distinguishing the address into three classes: Class A, Class B, and Class C. If users never plan to interface with the Internet, then they are free to use the IP address space any way they want. Otherwise, they must obtain official IP addresses—unique from others assigned—to operate over the Internet. You can obtain an official IP address through an Internet Service Provider (ISP) in your nearest metropolitan area. The Network Information Center (NIC) allocates IP network addresses and registers domain names. Network Solutions, Inc. was awarded the contract for NIC registration services in the early 1990s. The service providers coordinate address assignment with Network Solutions. For an official IP address, you will be given a unique network number and be free to assign addresses within the domain. For example, if you are assigned a Class C address, you will be free to assign up to 255 addresses.

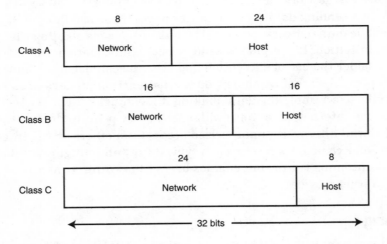

Figure 8.12
The Internet Protocol address hierarchy.

Address planning requires you to obtain enough official unique IP addresses for each network connection. For example, an organization having 350 users (each needing a network connection), 10 printers, and 4 servers would require at least 364 addresses. You could satisfy this requirement by obtaining one Class B address or two Class C addresses.

> **NOTE**
>
> Because of the vast number of organizations deploying Web servers and gaining access to the Internet, unique IP addresses are quickly running out. In fact, it's impossible to obtain a Class A address and very difficult, if not impossible, to obtain a Class B address. You will probably need to be issued multiple Class C addresses. The problem, though, is that it is difficult to manage multiple Class C addresses if they are not contiguous. Therefore, you will need to predict the number of addresses needed in the future to obtain a contiguous series of addresses.

If your network implementation requires a large number of IP addresses, or it is difficult to predict the number of addresses needed in the future, consider private addressing. The NIC has set aside a single Class A address that you can use within your network, giving you a large number of addresses. You have to agree, however, to not use these addresses on the Internet. If you need Internet access, you can deploy a proxy server that translates your private addresses into legal Internet addresses. This means you need to obtain at least one Class C address to support the connections to the Internet.

The second task of address planning is to assign a part of the address (subnet number) to each LAN connected to a router port. This address will be configured within the router using a subnet mask. You will then need to assign a unique address, within the applicable subnet number, to each network connection.

For smaller LANs (less than 50 users), you can install the IP address within the applicable configuration files at each PC. For large LANs, consider the use of DHCP (Dynamic Host Configuration Protocol). DHCP issues IP addresses automatically within a specified range to devices such as PCs when they are first powered on. The device retains the use of the IP address for a specific license period that the system administrator can define. DHCP is available as part of the Microsoft Windows NT Server NOS, and offers these advantages over manual installation:

○ **Efficient implementation of address assignments.** With DHCP, there is no need to manually install or change IP addresses at every client workstation during initial installation or when a workstation is moved from one location to another. This saves time during installation and eliminates mistakes when allocating addresses (such as duplicate addresses or addresses having the wrong subnet number). It is very difficult to track down address-related problems that may occur when using permanently assigned addresses. This advantage is greatest for larger networks.

○ **Central point of address management.** With DHCP, there is no need to update the IP address at each client workstation if making a change to the network's configuration or address plan. With DHCP, you can make these changes from a single point in the network. For example, if you move the domain name service software to a different platform, you need to incorporate that change for each client workstation if using permanently assigned addresses. With DHCP, you just update the configuration screen from a single point at the NT server or remote location. Again, this advantage is greatest for larger networks.

Network Management Protocols

In order to support effective network management, be sure to incorporate network management protocols into the design to support a network management system comprising the components shown in figure 8.13 and defined below.

○ **Network Elements.** Network elements, sometimes called managed devices, are hardware devices such as computers, routers, and terminal servers that are connected to networks.

○ **Agents.** Agents are software modules that reside in network elements. They collect and store management information, such as the number of error packets received by a network element.

○ **Management Information Base (MIB).** A MIB is a collection of managed objects residing in a virtual information store.

○ **Network Management Stations (NMSs).** An NMS executes management applications that monitor and control network elements.

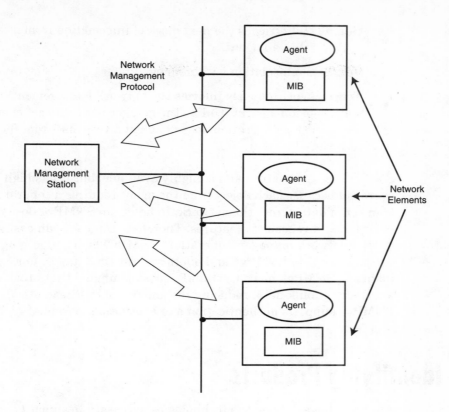

Figure 8.13
The network management system.

The industry's de facto standard for network management is Simple Network Management Protocol (SNMP), which facilitates the exchange of management information between network devices. By using SNMP to access management information data (such as packets per second and network error rates), network administrators can more easily manage network performance and find and solve network problems. If you plan to deploy network monitoring and control, ensure that the devices you want to manage support SNMP.

SNMPv1 is a simple request-response protocol offering the following operations:

○ **Get.** Retrieves a block of information from a table or list within the agent.

○ **Get-next.** Retrieves the next block of information from a table or list within an agent.

○ **Set.** Sets information within an agent.

○ **Trap.** Asynchronously informs the network management station of some event. Unlike the get, get-next, and set operations, the trap operation does not elicit a response from the receiver.

SNMPv1 is very mature and is highly available, but it has high overhead because each object from the agent requires a separate transfer of data across the network. In addition, SNMPv1 does not incorporate any security features. Therefore, someone can easily spoof a network management station. SNMPv2 is just emerging as a replacement to SNMPv1 and offers several advantages. For instance, SNMPv2 has a GetBulkRequest command that provides a transfer of bulk data, reducing the amount of overhead. Also, SNMPv2 supports authentication and encryption to combat spoofing.

Identifying Products

After you have defined the technologies necessary to support network requirements, you need to identify appropriate products. In some cases, such as the NOS and applications, you might have already selected the product as part of the network element definition phase. Regardless, select all products and materials necessary for implementing the network and create a bill of materials.

In general, select products based on the following criteria:

○ Ability to provide a necessary degree of functionality

○ Product availability

○ Level of vendor support after the purchase

○ Price

The following criteria are also important when selecting wireless products:

○ For wireless LANs, compliance with the IEEE 802.11 standard

○ Availability of tools that assist with installation (site survey tools, field strength meters, and so forth)

○ Availability of encryption for higher security

○ Ability to fit the form factors of your computers (ISA, PCMCIA, and so forth)

○ Ability to interoperate with the selected network operating system

N o t e

Appendix B, "Wireless Networking Products and Services," contains many of the wireless network products. You can use this as a guide for selecting wireless LAN, MAN, or WAN products.

Identifying the Location of Access Points

Most environments, such as hospitals, factories, and warehouses, cover an area exceeding the range of wireless LAN devices. As described in Chapter 2, wireless LAN vendors sell wireless local bridges (often called *access points*) to provide an interface between wireless users and wireline technologies such as IEEE 802.3 ethernet or IEEE 802.5 token ring. In addition, most of these access points also support wireless users who roam from one cell to another. A major design concern is identifying the location of these access points to provide an interface to network resources located on wired networks and to provide adequate coverage for roaming users throughout the facility.

It would be easy to deploy a wireless network in a completely open area, free from obstacles like walls, desks, and window blinds. This would allow radio waves from the wireless devices to maintain an omnidirectional radiation pattern, making it simpler to predict the maximum operating range between all devices and the locations of the access points. The presence of objects and the construction of the facility, however, causes attenuation to radio wave signals that distorts the radio propagation pattern, making it difficult, if not impossible, to predict. Table 8.1 gives you an idea of the degree of attenuation of various types of RF barriers.

Table 8.1

RF Attenuation for various types of barriers.

RF Barrier	Relative Degree of Attenuation	Examples
Air	Minimal	
Wood	Low	Office partitions
Plaster	Low	Inner walls
Synthetic Material	Low	Office partitions
Asbestos	Low	Ceilings
Glass	Low	Windows
Water	Medium	Damp wood, aquariums
Bricks	Medium	Inner and outer walls
Marble	Medium	Inner walls
Paper	High	Paper rolls
Concrete	High	Floors and outer walls
Bullet Proof Glass	High	Security booths
Metal	Very High	Desks, office partitions, re-enforced concrete, elevator shafts

When establishing requirements for the network, review the environment (see Chapter 6, "Defining Requirements for Wireless Networks") to give you enough information to select the appropriate type of wireless medium. To identify the location of access points, though, you need to evaluate the environment to a much higher level of detail, especially the effects of the environment on propagation of radio waves.

The best method for identifying the location of access points is to perform an RF site survey. Start by obtaining the following items:

○ Blueprints of the facility

○ At least one master station and NIC (PCMCIA form factor) from the selected vendor

○ Portable computer (the smaller the better) having a PCMCIA slot suitable for the wireless NIC

You should do the following when performing the site survey:

1. Verify the accuracy of the facility blueprints. Initial blueprints are drawn by an architect before the building is constructed. Changes are not always made to the drawings as the building is modified, especially for the relocation of walls and office partitions. You should walk through the facility before running tests to be sure walls are where they are supposed to be—if not, update the drawings.

2. Mark permanent user locations. On the blueprints, mark the location of users who will be operating from a fixed location.

3. Mark potential user roaming areas. In addition to the permanent users, outline potential user roaming areas within the building. In some cases, the roaming areas may be the entire facility. However, there may be some areas where users will never roam.

4. Identify obstacles that may offer significant attenuation to the radio waves. Observe the construction of the facility, and mark the location of obstacles that may cause a hindrance to radio wave propagation.

5. Identify potential sources of interference. You should have done this when reviewing the environment as part of the definition of requirements. If not, determine what other RF devices are present, and assess their effect on the wireless LAN. You can do this by talking to someone at the facility who manages existing RF devices, or you can use a spectrum analyzer to record RF transmissions that fall within the frequency band your wireless LAN will operate. Outline the areas on the blueprints that the sources of interference will affect.

    This step is important! A company in Washington D.C. purchased 200 wireless LAN cards, installed them, and later discovered radio interference from a nearby military base blocked the operation of half the users. A proper verification of coverage or even the use of a spectrum analyzer would have saved this company a great deal of money and frustration.

6. Identify the preliminary location of access points. Based on the wireless LAN vendor's range specifications and information gained from steps 4 and 5, identify the preliminary location(s) of access points and wireless servers. The goal is to ensure all permanent and roaming users can maintain access to applicable network resources via access points. For small areas (less than 1,000 feet in diameter), no access points may be necessary; however, larger areas will require access points to produce a multi-cell system. Mark the presumed locations on the blueprints.

7. Verify the location of access points. This is best done by installing an access point or master station at each of the locations identified in step 6, and then testing the signal strengths at each of the permanent and potential user locations. You need to configure a portable computer with the applicable wireless LAN adapter and site survey software supplied by the vendor. Proxim, for example, ships a site survey tool with their RangeLAN products that loads on the portable computer and broadcasts messages to all other units with the specified domain. Each unit responds to these broadcasts, and the survey tool, after ten broadcasts, displays a Link Quality number that represents the percentage of packets to which the tool received a response. The tool displays a

Link Quality of 5 for 100% packet acknowledgment. A Link Quality of 4 represents 80%, Link Quality of 3 represents 60%, and so on. A Link Quality of 5 is optimum, but operation is still possible at lower quality levels. (Users will experience some delay though.). With the appropriate tool loaded, walk with the portable computer and record the signal qualities at all applicable locations. If the signal quality falls below suggested values supplied by the vendor, then consider relocating the access point.

Verifying the Design

Design verification ensures that the solution you have chosen will support requirements. Actually, the verification of access points covered in the preceding section was a form of design verification that tested the wireless network portion of a network architecture—the physical and data link layers. But you might also need to verify higher layer architectural elements, such as applications, communications protocols, and system interfaces as well. This prevents the purchase of inappropriate network components and hours spent working out bugs just before the system needs to be operational.

The following are methods you can use to verify the design:

○ Physical prototyping

○ Simulation

○ Design review

Verifying the Design Through Physical Prototyping

Construction and testing of the part of the system you want to verify is known as *physical prototyping*. It uses the actual hardware and software you might eventually deploy. In some cases, you could include the prototype as part of the initial implementation, perhaps as a system pilot. The prototyping can also take place in a laboratory setting or testbed.

The following are the main attributes of physical prototyping:

- ○ Yields very accurate (real) results because you are using the actual hardware and software

- ○ Is relatively inexpensive because you can obtain components under evaluation from vendors

- ○ Takes time to reconfigure

- ○ Requires access to network components, which can be a problem if you do not have easy access to vendors (from remote areas, for example, such as ships at sea, the South Pole, and so forth)

- ○ Requires space to layout the hardware

Typically, you do not need to physically prototype the entire system, especially those parts that other organizations have implemented without encountering problems. Consider prototyping any solutions that have not been tested before. The following are some examples:

- ○ Interfaces between wireless users and network resources located on a wireline network

- ○ Access from users to mainframe applications

- ○ Operation of newly developed applications

- ○ Operation of the system in areas where there is a high potential for inward and outward interference

Verifying the Design through Simulation

Simulation is software that artificially represents the network's hardware, software, traffic flows, and utilization as a software model. A simulation model consists of a software program written in a simulation language. You can run the simulations and check results quickly, greatly compressing time by representing days of network activity in minutes of simulation runtime.

The following are the main attributes of using simulation for verifying the design:

- ○ Results are only as accurate as the model. In many cases you need to estimate traffic flows and utilization.

○ After building the initial model, you can easily make changes and rerun tests.

○ A simulation does not require access to network hardware and software.

○ A simulation does not require much geographical space, just the space for the hardware running the simulation software.

○ Simulation software is fairly expensive, making simulation economically unfeasible for most one-time designs.

○ The people working with the simulation program will probably need training.

For most implementations, you do not need to run simulations. Consider using simulation for the following situations:

○ When needing a better understanding of the bandwidth requirements (system sizing) based on predicted user activity. (It is not practical to do this with physical prototyping.)

○ If you are in an area where it is difficult to obtain hardware and software for testing purposes.

There are simulation tools on the market that can assist designers in developing a simulation model. Most simulation tools represent the network using a combination of processing elements, transfer devices, and storage devices.

Mil 3's Opnet

Mil 3's Opnet simulator has evolved as a response to the problems resulting from network complexity. Structured around a top-down, graphical hierarchy of representation that uses the latest software technology, nodes are represented as objects that communicate through data-flow networks and can be quickly customized to specific details. The entire system, including statistical analysis of network traffic, is portrayed through an X Window graphical user interface. As a result, the designer can think in terms of basic architectures and explore the consequences without coding a design from the bottom up.

Opnet is structured as a series of hierarchical graphical editors that address each level of network design. Consisting of three levels, the highest tier is based on connectivity, operating as a schematic-capture function. Graphical representations of a network can be superimposed on backgrounds representing floor plans or geographic areas. The second level, the node editor, captures node activity in terms of data-flow analysis of hardware and software subsystems. The third level contains a process editor that defines the control flow, such as a protocol or algorithm.

American Hytech's NetGuru Simulator

American Hytech's NetGuru Simulator is a fully functional network simulation module that is integrated with NetGuru Designer. The Simulator uses Object Oriented Programming techniques with graphical icons that enable even nontechnical users to perform design and simulation activities. The NetGuru family of tools have plug-and-play Microsoft Windows-based modules that represent generic physical and logical network elements. The product addresses the needs of professionals dealing with LANs and hundreds of workstations, or larger networks that can be further segmented.

NetGuru provides the ability to design the network from scratch or document an existing design and conduct a "what if" analysis. This helps the designer understand whether the networks are performing efficiently, identify potential bottlenecks, and predict the effect of new hardware, software, users, network configurations, or network tools as they relate to overall performance.

Verifying the Design through Design Reviews

Whether you have performed simulation and physical prototyping or not, you should conduct a design review as a final verification process. This review ensures there are no design defects or issues before pressing on with component procurements. It is best to have the entire team, especially analysts and engineers, review the design to ensure that it will adequately support all requirements. Analysts should raise questions related to the ability of the design to satisfy requirements, and engineers should be able to fully explain how the design will meet requirements. For first-time complex implementations, consider hiring a consultant to verify the design. This could eliminate many problems when installing and testing the system.

During the design review, participants should:

1. Review all design documentation.

2. Identify defects in the design.

3. Describe potential technical problems.

4. Recommend further prototyping or simulation of unclear design specifications.

Be sure to use lessons learned from other projects to spot problems in the inability of the specifications to meet requirements.

Documenting the Final Design

As with requirements, you need to document the details of the design to support further implementation activities, such as component procurements, installation, and so forth. Final design documentation should include the following:

○ A description of each network element

○ The location of access points

○ Standards

○ The products necessary for satisfying specific requirements

TIP

Be certain to update any documentation prepared throughout the design with any changes made after verifying the design. Also, update the project documentation, such as the budget, schedule, and resources required to complete the project.

The last step before procuring the components is to obtain approvals for the design. This ensures that applicable managers agree to fund the implementation shown in the design. Those involved usually include network configuration management, the customer representative, and people with funding authority. For approvals, you can have these people sign a letter with at least the elements shown in figure 8.14. After the approval, consider the design as a baseline that can only be changed by following the stated change control procedures.

Approval Letter

- Design document number
- Change control procedures
- Sign offs
  - Technical manager
  - Project manager
  - Customer Representative
  - Funding Authority

Figure 8.14
The major elements of an approval letter.

Procuring Components

When procuring components, you need to understand the warranties and maintenance agreements vendors offer. Most vendors offer excellent warranties and also have maintenance agreements at an additional charge.

You should ask vendors the following questions:

○ How long is the product covered?

○ What are the limitations of the coverage?

○ How should the product be returned if it becomes defective?

○ Does the vendor provide on- or off-site maintenance?

Before ordering the components, you should plan where the components will be stored after delivery. For small implementations, this may not be significant, but for larger implementations, it is crucial. For example, imagine ordering 75 PCs, 150 network interface boards, and 5 printers. Do you know where to put all the boxes when they arrive? Because implementations of this size or larger require a great deal of space to store components before they are needed for installation, plan the following items:

○ The location to which the components will be delivered.

○ Storage locations for components waiting for installation.

○ The mechanisms for moving components from the delivery point to the storage area.

○ The mechanisms for moving the components from the storage area to the point of installation.

Planning the Support of a Wireless Network

Before turning the new system over to users, you need to make sure you've planned and prepared for how you're going to support the network. The goal with this phase of the project, which can be accomplished in parallel with the design phase, is to make certain the system continues to operate effectively during its production phase. This chapter discusses the following elements of support you should prepare prior to implementing the system:

○ **Training.** Training provides users, system administrators, and maintenance staff the know-how to effectively operate and support the new system.

○ **System administration.** System administration is the liaison between the system and its users. With a network, a system administrator manages the network operating system.

○ **Help desk.** The help desk is a central point of contact for users needing assistance with utilizing the network and its resources.

○ **Network management.** Network management provides a variety of elements that protect the network from disruption and provide proactive control of the configuration of the network.

○ **Maintenance.** Maintenance staff members perform preventative maintenance on the network and troubleshoot and repair the network if it becomes inoperable.

○ **Engineering.** Engineering assists system administrators, the help desk, and maintenance staff in troubleshooting difficult network problems.

○ **Configuration control.** Configuration control procedures make certain proper control procedures exist for making future network changes.

Training

Training is an important aspect of any activity, from flying the space shuttle to changing baby diapers. With networks, users need to know how to access network resources and run applications, and the system administrator needs to understand how to manage the network operating system. The implementation of proper training significantly increases the effectiveness of a new system because users have less of a learning curve, minimizing the drop in productivity normally encountered with new systems. Also, the users require less support from system administrators and the help desk.

When preparing for the delivery of training, perform these two steps:

1. Determine training requirements.

2. Create course materials.

The following sections explain each of these steps.

Determining Training Requirements

The analysis of training requirements is similar to defining requirements for the overall system—you determine *what* people need to learn. Here are some steps you should perform when determining training requirements:

1. Interview a sample of potential students.

2. Develop a draft list of learning objectives.

3. Develop a training plan.

4. Review, edit, and approve the training plan.

As with the initial system requirements determination phase, you should interview a representative sample of potential students to determine learning objectives for that particular group of users.

These learning objectives identify what the student is supposed to know after receiving instruction. For user training on new applications, be sure to ask questions related to the students' experience with other applications and what parts of the new application they will be using. For system administrators, you should determine what experience and past training they've had with the chosen network operating system.

After interviewing the students, list the learning objectives of the users, system administrators, and any other support staff requiring training. In general, users need to know how to log in to the network operating system and run applications. System administrators need to learn how to set up user accounts, configure printers, run backups, execute disaster recovery procedures, and perform troubleshooting.

Part of determining training requirements is developing a training plan that indicates, as shown in figure 9.1, the learning objectives, course descriptions, and the schedule of course offerings, as well as the source and location of training. Be sure to have upper management and/or the customer representative review the training plan before proceeding with the acquisition/development and delivery of the training courses.

```
┌─────────────────────────────────┐
│        Training Plan            │
├─────────────────────────────────┤
│                                 │
│  • Learning Objectives          │
│                                 │
│  • Course Descriptions          │
│                                 │
│  • Schedule of Course           │
│    Offerings                    │
│                                 │
│  • Training Locations           │
│                                 │
└─────────────────────────────────┘
```

Figure 9.1
The elements of a training plan.

Generally, it is cheaper and easier to acquire off-the-shelf training from training vendors. Many training companies, for example, offer system administrator training on Novell NetWare and Microsoft NT Server at less expense than developing and delivering the courses internally (unless you're training a thousand people and have an effective training staff). You may need to develop your own custom courses, however, preferably through someone with the right experience, for new applications.

Creating Course Materials

If you decide to develop your own training program, you need to perform the following tasks to create course materials:

1. Develop lessons, exercises, and visual aids for each course.

2. Identify and develop precourse materials.

3. Create an instructor guide for each course.

4. Develop student course books.

5. Instruct an initial pilot offering of each course.

6. Modify the course if necessary.

The development of effective training courses requires someone who understands the learning objectives. It is also beneficial if the course developers are able to teach the course as well. This reduces the amount of effort in developing the instructor guide and shortens the preparation time of the instructors because they have first-hand knowledge of the course materials.

The following describes each of the training elements you need to create:

○ **Lessons.** Lessons describe what the instructor should teach. Each lesson should correspond to the first or second level of the course outline. Like a chapter of a book, lessons should contain related items. There should be a combination of lessons comprising each course that consists of the following:

  ○ Lesson overview identifying learning objectives.

  ○ Concepts and procedures the instructor should explain and demonstrate.

  ○ Expected level of skill the students should acquire.

  ○ Method of instruction.

The method of instruction should be one or more of the following:

Lecture. A lecture is primarily a one-way information flow from the instructor to the students and is efficient for conveying information to large groups. Lectures are quite familiar—it is the primary form of instruction from elementary school through college. Lecture is the most common form of instruction. If used alone, however, it is the least effective.

Discussion. Discussion is a uniform information flow among students and the instructor. A well-managed discussion within the classroom is good for exchanging ideas among a small group, especially if some students have experience in the subject area. Incorporate discussions in the course, if possible, to add variety.

Exercises. In doing exercises, the student actually uses the application or performs some procedure. Generally, the instructor should give some background and demonstrate the procedure before having the students perform the exercise. Exercises are the most effective form of learning in a classroom. People always retain more information if they perform as they learn.

TIP

Be sure to utilize a good balance of instruction types throughout the course. A lecture-only course will generally put students to sleep. Add some discussions and exercises to liven things up.

○ **Exercises.** As previously mentioned, exercises add much value to training courses. When developing exercises, be sure to tie them to learning objectives and write clear instructions for students to follow. It is always a good idea for the course developer to have someone practice the exercises before administering them to the students.

○ **Visual Aids.** Visual aids are items students see that enhance the delivery of the course. Examples of visual aids include transparencies, 35mm slides, pictures, screen shots, and videotapes. People will grasp concepts much easier by observing good visual aids ("a picture is worth a thousand words"). For transparencies, use concise bullet statements (preferably five words or less each), avoid more than four or five bullets per slide, and use a font style and size that is easy to read when projected on the screen. The use of different colors and graphics can enhance the charts, but be careful of overdoing it.

○ **Precourse Materials.** If you need to compress the length of the courses, consider the use of precourse materials, such as articles or books the students should read before attending the course.

○ **Instructor Guide.** An instructor guide explains how to teach a particular course. This is especially useful if the instructors are people who were not involved in the course development. You can create an instructor guide by clearly describing the lessons in a standard format. Figure 9.2 identifies the main elements of an instructor guide.

```
┌─────────────────────────────┐
│      Instructor Guide        │
├─────────────────────────────┤
│                              │
│  • Course Learning Objectives│
│                              │
│  • Course Description        │
│                              │
│  • Course Audience           │
│                              │
│  • Lessons                   │
│                              │
│  • Identification of         │
│    Visual Aids               │
│  • Course Preparation        │
│    Procedures                │
│                              │
└─────────────────────────────┘
```

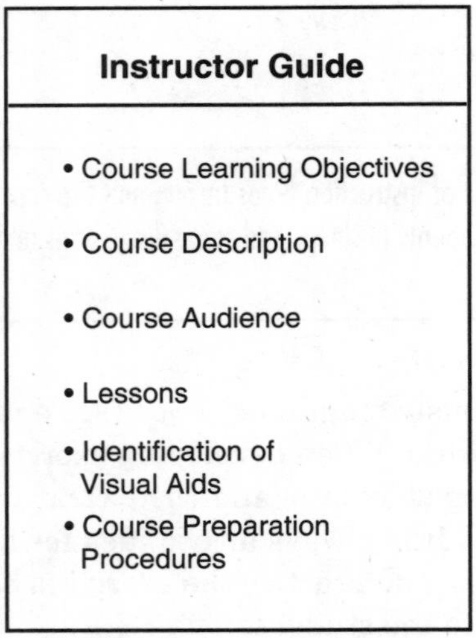

Figure 9.2
The main elements of an instructor guide.

○ **Student course books.** Student course books should include useful items the students need during class or reference material they can use after attending the course. The books should contain background information, procedures, exercises, glossary, and a place for all handouts.

○ **Initial pilot offering.** A pilot offering of a newly developed course is good to smooth out wrinkles. The pilot offering is also a good time to introduce the course to future instructors. Be sure to include enough time after the pilot offering to make necessary course modifications before offering the course again.

System Administration

If the system you are deploying consists of a network operating system or applications requiring system administration, you need to assign someone to be the system administrator. For smaller implementations (less than 30 users), this person could act part-time as the administrator and perform other tasks as well. Larger networks usually require a full-time administrator. In either case, be certain the administrator either has experience in the applicable systems or that he will be fully trained.

After you have identified a system administrator, prepare to incorporate the new system by doing the following:

1. Assign user names and passwords.

2. Assign access rights.

3. Plan directory structures.

Assigning User Names and Passwords

You should identify who the initial users of the network will be and determine their account names and initial passwords. You can obtain the names of users from the definition of requirements or by obtaining employee listings from human resources. Be sure to standardize user account names to increase efficiency in assigning user names. This simplifies the association between the user name and the actual user. For instance, establish user names with first and second initials followed by the last name. If two users have similar names, one of the user names could be followed by a number.

For example:

> John C. Jones becomes JCJONES
>
> Jack A. Roberts becomes JAROBERTS
>
> Julie A. Roberts becomes JAROBERTS2
>
> Santa Claus becomes SCLAUS

Assign an initial password to each user (such as their user name) and then insist the users change the password immediately after logging in to the system for the first time. The length and format of passwords should satisfy all security requirements and applicable regulations and policies. Typically, the length of the password should be at least six characters long, and the format ought to consist of alphanumeric characters. The system administrator should make certain users change their password periodically to avoid compromising password security.

Also think about associating sets of related users as groups. All users within Accounting, for example, could be given the group name ACCOUNT. This makes it easier to assign common rights to particular departments.

Assign meaningful names to servers and printers within the network. For example, utilize a name such as APPS1 as a name for a server containing applications. Avoid using names that associate the server to its platform type and location, which is more likely to change than the server's function.

Assigning Access Rights

For each user and group, assign the proper rights to specific servers, applications, directories, and files. As a default, network operating systems assign all rights to all users. You need to change the rights assignments to limit users' access to resources based upon job responsibilities. Review security requirements and assign rights to users based on their need to access certain information.

Typically, you can assign a combination of the following rights to each user and group:

○ **Supervisory**—full rights and ability to assign rights.

○ **Read**—ability to read directories and files.

○ **Write**—ability to write to files.

○ **Create**—ability to create directories and files.

○ **Erase**—ability to erase directories and files.

○ **Modify**—ability to rename files and change file attributes.

○ **File Scan**—ability to see directories and file names.

○ **Access Control**—ability to assign rights for a particular directory.

The proper issuance of these rights constrains users to the specific directories, files, and applications they need to access.

Planning Directory Structures

A directory structure provides a framework for the installation of applications and storage of files. Start by establishing a set of drives containing common types of applications and files as follows:

Drive	Purpose
A–D:	Local client drives (disks, hard drives, CD-ROM)
E:	Network operating system utilities
F:	Users' personal directories for storage of files
G:	Program specific directories
H:	Shared public drive (for storage of company and project information, and so forth)
I–P:	Assigned as needed by users
Q–Z:	Reserved

Then continue by assigning subdirectories to the drives as needed. For example, you could assign a subdirectory on the F: drive for each user as follows:

F:\JAROBERTS

F:\JAROBERTS2

F:\JCJONES

F:\SCLAUS

And, the G: drive could have directories as follows for each application program:

G:\WORD (for Microsoft Word)

G:\EXCEL (for Microsoft Excel)

Help Desk

The help desk is the central point of contact for users needing help with an application or other system-related problem. The help desk performs functions and operations as shown in figure 9.3. For instance, a user having a question or problem calls a single phone number and reaches the help desk. The help desk staff should attempt to answer the user's questions or solve the problem over the phone if possible, minimizing the amount of time the user has to wait. In some cases, the user may not be able to perform his tasks until the problem is fixed. If the help desk staff cannot satisfy the user's request, then the help desk person should classify the problem and hand it off to a second level of support—such as maintenance, engineering, or an outside consultant. For network problems, the help desk should initiate and manage a trouble ticket until the problem is resolved. The trouble ticket describes the problem and the status of the repair.

The help desk should concentrate on user satisfaction. Here are some suggestions when establishing or upgrading a help desk:

- Develop clear procedures outlining how the help desk operates.
- Establish a single phone number (with multiple call handling) for the help desk.
- Ensure all users know the phone number of the help desk.
- Plan for increased call volume as the network grows.
- Incorporate a method to effectively track problems.
- Fully train the help desk team in network operations principles, particularly in user applications.

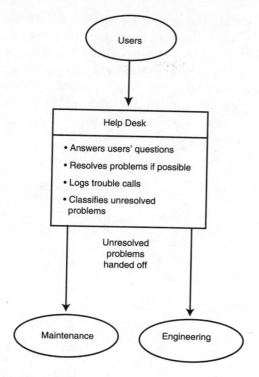

Figure 9.3
The functions and operation of a help desk.

○ Use surveys to determine user satisfaction with the help desk.

○ Incorporate methods to reduce the stress of the help desk staff.

○ Review help desk usage statistics to determine optimum staffing.

○ Periodically rotate network implementation and system administration people into the help desk.

Network Management

Network management enables you to evaluate, plan, and control the configuration of your network. Support staff can then effectively control the network and respond faster to problems. Network management is also responsible for high-level troubleshooting of network problems, at least to the level of detail that the network management software allows. Network management software, for example, typically enables you to isolate problems to the component level, such as a cable, switch, or network interface card. Generally, maintenance people are required to drag test equipment out to the problem location to perform troubleshooting within these components. Figure 9.4 illustrates the elements of network management.

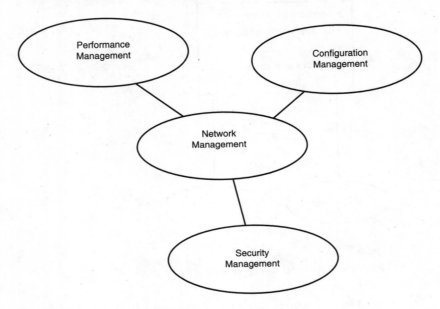

Figure 9.4
The elements of network management.

The following sections describe each of the network management elements:

Performance Management

The goal of performance management is to make certain the network meets the performance requirements of the users. Performance management includes performance modeling and monitoring that measures elements, such as network throughput, user response times, and line utilization, to facilitate proactive control of the network.

Performance Modeling

Performance modeling is the use of simulation software to predict network behavior, enabling you to perform capacity planning. With simulation, you can model the network and impose varying levels of utilization to observe the effects. A necessary component of performance modeling, though, is performance metrics, such as user activity levels. Accurate performance metrics are important because they provide a basis for the simulation; however, you must often predict metric values, increasing the potential for error. Consider the use of performance modeling on large, complex, or mission critical networks. OPNET, which is described in Chapter 8, "Designing a Wireless Network," is a fully featured network simulation tool. Consider its use for simulating the performance of mission-critical networks.

Performance Monitoring

Performance monitoring addresses performance of a network during normal operations.

You should concentrate on establishing a means to monitor central points of failure and any part of the network reliant on mechanical devices. Central points of failure tend to be components such as servers, network backbone cabling, switches, routers, and access points.

The three main types of performance monitoring are as follows:

○ **Real-time monitoring**—where metrics are collected and compared against thresholds that can set off alarms.

○ **Recent-past monitoring**—where metrics are collected and analyzed for trends that may lead to performance problems.

○ **Historical data analysis**—where metrics are collected and stored for later analysis.

A network monitoring station identifies data traffic and collects statistics from workstations, servers, and other active network devices. These monitoring stations depend on the implementation of a management protocol, providing communications between the network monitoring stations and the managed devices. As mentioned previously, SNMP is currently the industry standard. Network monitors attach to the network via a network interface card and run in a passive mode, unable to disturb the normal operation of the network.

Network monitors usually consist of the following features:

○ Graphical mapping of the network components and environment

○ Event logging

○ Alarms triggered when certain events occur

○ Automatic fault isolation

○ Statistics logging

○ Report generation

Plan to utilize network management tools that interface with the MIBs residing on the network's active elements. Most wireless LAN vendors, for example, supply a SNMP-based MIB that can be read by most network management software. Lucent's WaveLAN comes with a MIB that interfaces with Novell's NetWare Management System. For monitoring switches and hubs, you can utilize software sold by the vendor because it will interface best with the MIBs located on the devices; that is, if you are using 3Com's switches and hubs (their network monitoring product), Transcend would be the best choice.

Cabletron's SPECTRUM

If your network implementation consists of a variety of vendors, consider using a common network monitoring product such as Cabletron's SPECTRUM.

SPECTRUM Version 4.0 has the following features:

○ Provides advanced management capabilities for all networking environments—LAN, WAN, SNA, PBX and ATM—and all computing environments.

○ Proactively monitors the network's or computer's current status and performance characteristics.

○ Provides intelligent alarm reduction to help minimize time required to locate faults.

○ Automatically isolates both hard and soft errors.

○ Takes corrective action to assist network personnel in solving problems.

○ Discovers configuration information, providing a broader perspective of the network.

○ Collects and analyzes valuable management data for short- and long-term network planning.

○ Multiple platform support (Windows NT, Solaris 2.4, SunOS 5.4, SGI Irix 5.3, HP-UX 10.01, AIX 4.1.3, and SunOS 4.1.3).

○ Enterprise Alarm Manager provides a view of all alarms in entire management system, enterprise-wise applications.

○ Alarm Notification Manager dispatches SPECTRUM alarms to external applications like trouble-ticketing packages

○ AutoDiscovery maps your network by identifying IP addresses and placing devices in a logical hierarchy. As devices are added to your network, AutoDiscovery automatically updates your network model.

○ MAC Address Locator Tool (MALT) locates devices on your network when you know the Physical or Media Access Control (MAC) address.

Configuration Management

The goal of configuration management is to track the versions of hardware and software elements residing on the network. When preparing for configuration management, decide what elements you want to manage the configuration of and select the appropriate tools. You should track the versions of software located on PCs and network operating systems, as well as the settings on switches and hubs. Most network operating system vendors include configuration management software as part of their network management solution. For example, you can use Microsoft's SMS to query MIBs located on network interface cards, servers, and other elements to retrieve listings of software versions and configurations. You can set up the network management software to periodically obtain the information and store it in a database for future reference when troubleshooting problems.

Microsoft's System Management Server (SMS) 1.1

Microsoft's System Management Server (SMS) 1.1, which is designed to run on Microsoft Windows NT Server, primarily manages client and server software and hardware—devices capable of running the SMS client software element. SMS maintains a centralized Microsoft SQL database that stores configuration information of the SMS-managed devices. SMS detects the presence of machines and keeps an inventory of elements, such as applications, processor, operating systems, adapters, and available disk storage space. SMS also enables centralized distribution of software to networked clients. If upgrading an application, for example, you can install the new version on everyone's computer via SMS.

Security Management

Network security protects information and ensures nothing interferes with operation of the system. The main idea is to take precautions to reduce the chance of security-related incidents and develop procedures for countering if something bad does occur. Network security mechanisms consist primarily of access control and the planning and execution of disaster recovery procedures. When preparing for security management, assign user passwords and resource rights as explained earlier in this chapter. In addition, be sure to incorporate virus protection procedures and plans for recovering from disasters.

Protecting Against Viruses

Viruses can enter your system via files downloaded from external networks and computers. The best way to protect against viruses is to follow the procedure illustrated in figure 9.5. Be sure to control viruses by scanning any files being loaded onto the system. Also, periodically check for the presence of viruses on the network as well.

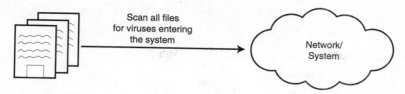

Figure 9.5
Virus protection procedures.

Norton's AntiVirus

Norton's AntiVirus is an example of AntiVirus software. AntiVirus installs in your system and works automatically in the background, watching for disk activity that could indicate a virus. Norton AntiVirus detects and eliminates many types of viruses, ensuring backups and file transfers are virus-free.

Planning for Disaster Recovery

A disaster situation is anything that disrupts or has great potential of disrupting the use of the network. The detection of a virus on a PC would constitute a crisis situation because the virus may potentially spread to other PCs and cripple network operations, which could lead to a disaster. An electrical power blackout, accidental file deletion, and widespread virus attacks would most likely constitute a high level disaster in that they would cause a substantial disruption of the system.

To enable a swift reaction to disasters, develop disaster recovery plans offering step-by-step procedures on what to do if the network goes down. Generally in a panic situation, people will not be able to think clearly enough to react quickly. If you were flying an airplane and the engine went out, what would you do? Hopefully you wouldn't become hysterical, but a well-tested checklist would surely help calm your nerves. A network disaster is not always a life-and-death situation, but there can be enormous pressure on the system administrator when a server's hard drive crashes. You'd better play it safe and prepare a disaster plan having at least the major components shown in Figure 9.6.

```
┌─────────────────────────────────────┐
│            Disaster Plan             │
├─────────────────────────────────────┤
│                                      │
│  Description of                      │
│  • Potential Disasters               │
│       • Power outage                 │
│       • Flooding                     │
│       • Theft                        │
│       • Server crash                 │
│       • Wind damage to outdoor       │
│         antennas                     │
│  • Disaster recovery procedures      │
│       • Who should implement the     │
│         procedures                   │
│       • Step-by-step instructions    │
│                                      │
└─────────────────────────────────────┘
```

Figure 9.6
The major components of a disaster plan.

During the operation of the network, occasionally run drills, if feasible, to test the disaster plans.

Maintenance

To perform preventative maintenance and repair network problems, you should establish a field service team that includes technicians who can troubleshoot network problems. Preventative maintenance helps spot network problems before they impact users. What type of preventative maintenance is necessary for networks? The general answer is anything that checks the condition of mechanical elements, such as cabling, connectors, and antenna positioning. In addition, for radio-based wireless networks, be sure to periodically check for potentially interfering radio signals. When you initially defined requirements and performed the site survey, any signals or other radio devices may not have been present that could have affected the LAN's performance. However, as time passes, the organization or a nearby company could install a system that could cause interference problems. A periodic spectrum analysis or a simplified site survey would discover this before a problem occurs.

To adequately perform maintenance, whether it's a preventative or reactive response to problems, you need to obtain some test equipment, such as a spectrum analyzer and protocol analyzer. As mentioned before, a spectrum analyzer will display the amplitude of signals present at various frequency points of interest. Therefore, you can use the spectrum analyzer to determine whether there are any signals present that may affect the wireless LAN propagation.

A protocol analyzer enables technicians to visualize the operation of network protocols as they traverse the network. A protocol analyzer captures data traffic, decodes and displays protocol headers and data, traps specific protocol functions, generates and transmits test data traffic, and triggers programmed alarms. These features enable technicians to efficiently troubleshoot network problems.

Engineering

To provide support of difficult problems, particularly those that may lead to a network modification, have an engineering staff either in-house or available on a contract basis. The engineering staff should maintain knowledge of the inner workings of all network components and stay abreast of applicable network technologies.

Problems will normally come to the engineering group from maintenance or, in some cases, the help desk (see fig. 9.7). A user named Sally within a warehouse may call the help desk identifying that she is experiencing a great deal of delay accessing the inventory database from her wireless terminal. The help desk may hand the trouble call off to network management who runs tests and doesn't find any network problems that could attribute to her delays. Network management may then hand the trouble call over to maintenance to do some lower level troubleshooting. For this case, maintenance runs some spectrum analysis checks and finds the presence of a periodic interfering radio signal. They hand the problem over to engineering to find a solution, which may require the repositioning of an access point.

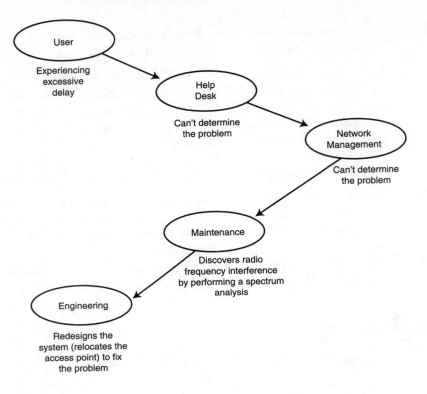

Figure 9.7
Problem progression from a user to the engineering staff.

The previous example describes the reactive function of the engineering staff; however, engineering can play a significant proactive role as well by implementing a network reengineering methodology. Network reengineering, as shown in figure 9.8, constantly monitors factors that may influence you to modify the network, ensuring the network remains tuned to best meet the needs of the users.

Factors that may influence changes to the network include:

○ Company relocations

○ Reorganizations

○ New applications and technologies

To perform modifications, you can use the overall network implementation process described in this book as a basis (see Chapters 5–10).

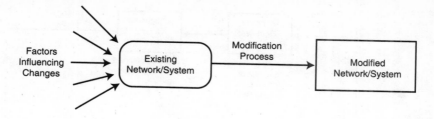

Figure 9.8
The network reengineering process.

Configuration Control

If nothing ever changed, life would be pretty boring. However, changes to the network, especially those not managed, can cause a great deal of headaches. The lack of proper control over changes to the network can result in stovepiping, where different portions of the company don't coordinate with each other or anyone else for that matter and deploy systems and applications that are not interoperable. Providing interfaces that allow systems and users at these dissimilar sites to share information then becomes difficult, expensive, and perhaps impossible. Also, the lack of control over network implementations makes it difficult and costly to support the systems. For instance, you may end up with three different types of network operating systems and four different types of wireless LAN adapters to support. Centralized support would need to keep abreast of all of these product types, resulting in higher training costs.

As you prepare for the operational support of a company-wide system, establish a configuration control process as shown in figure 9.9. The design and installation of the system consists of hardware, software, documentation, procedures, and people. It's paramount to consider the system implementation as a baseline, to be changed only when the person initiating the change follows the process. To implement this process, though, you need to identify those elements (configuration items) that are important to control.

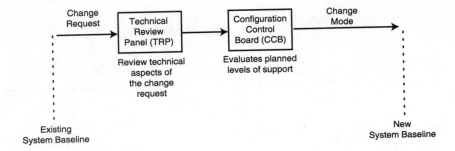

Figure 9.9
The configuration control process.

The following identifies examples of configuration items you should consider incorporating as a basis for the configuration control process:

- ○ Network interface adapter vendor
- ○ Network operating system release
- ○ Cabling standard
- ○ Switch vendor and type
- ○ Support plans

The description of these elements should be stored in a library accessible by the entire organization.

The person wanting to make changes to configuration items must submit a change request to the technical review panel (TRP) which will assess the technical nature of the change. The change request includes an evaluation of whether the change complies with the company's technical standards. If the TRP feels the change is technically feasible, it forwards the request to the configuration control board (CCB) for final approval. The CCB mainly evaluates whether the project team has prepared for adequate levels of support for the implementation and coordinated the changes with the proper organizations. With approval of the change, the project team then must ensure the preparation of support documentation.

Documenting Plans for Operational Support

As with all phases of a project, documentation is important to convey the ideas from that phase to other project phases. In the case of operational support, a support plan is necessary to effectively carry out the support. An operational support plan describes how the organization will support the operational network. This plan should indicate which network elements require support and which organizations are going to support them; therefore, the plan should address all support elements this chapter describes. Figure 9.10 identifies the major elements of an operational support plan.

```
┌─────────────────────────────────────────────┐
│  Operational Support Plan                     │
├─────────────────────────────────────────────┤
│                                               │
│   • System/Network description                │
│     (reference to design documents)           │
│                                               │
│   • Definition of elements requiring          │
│     support                                   │
│                                               │
│   • Description of how support                │
│     will be accomplished                      │
│                                               │
│   • Identification of who will                │
│     provide support                           │
│                                               │
└─────────────────────────────────────────────┘
```

Figure 9.10
The major elements of an operational support plan.

Preparing for the Transfer to Operational Mode

The transfer of the system to operational mode should be very well defined. Otherwise, it's not clear who is supporting the network, and "finger pointing" will occur if any problems arise. The main task of preparing for the transfer to operational mode is to develop a turnover agreement, outlined in figure 9.11, that will be put into effect after completing the installation and testing phase. At the beginning of the project, the project charter gave the project manager responsibility for implementing the network. The turnover agreement transfers this responsibility to the supporting organizations, the handling system administration, network management, and so on.

Turnover Agreement

- Description of support
  (reference to the operational
  support plan)

- Endorsements:
  - Project Manager
  - Customer Representative
  - People responsible for
    supporting the network/
    system

Figure 9.11
The outline of a turnover agreement.

The steps taken in this chapter will make the transition to an operational network smoother. With operational support ready to go, you are set to begin the installation phase.

Installing a Wireless Network

Wireless networks are advantageous because less cable is installed than with wired networks such as ethernet. Theoretically, then, a wireless network installation should take less time. Actually, installation time depends on how well you have evaluated the environment before embarking on the installation. If you did not perform a site survey as explained in Chapter 8, "Designing a Wireless Network," unseen radio frequency interference might wreak havoc on the operation of a newly installed wireless system, causing significant delays in making the system operate effectively. You want to avoid delays when installing the system. A delay will cause you to not meet schedule constraints and will result in decreased productivity of the system's users. Thus if you have not performed a site survey at this point in the project, do it now before pressing on!

The installation of a wireless network requires the following steps:

1. Plan the installation.

2. Install the components.

3. Test the installation.

This chapter covers each of these steps and explains the actions necessary to finalize the project.

Planning the Installation

Before taking components out of the boxes, installing network interface cards, and setting up antennas, spend some time planning the installation. Planning will significantly reduce the number of problems that may arise. Planning the installation consists of the following actions:

1. Developing an installation plan

2. Coordinating the installation

Developing an Installation Plan

Overall, an installation plan explains how to install the network. Developing an installation plan helps you focus on what needs to be installed. It also provides instructions for installers who might not have been involved with the design of the network and, therefore, do not have first-hand knowledge of the network's configuration. Figure 10.1 identifies the major components of a network installation plan. The project team should assign someone as installation manager who will develop the plan and be responsible for the installation.

Installation Plan

- Points of Contact
- Safety Tips
- Installation Procedures
- Tools
- Reference to Design Documentation
- Schedule
- Resources
- Budget
- Risks

Figure 10.1
The contents of an installation plan.

Points of Contact

The plan should indicate someone as the central point of contact if issues arise. This person should work in the facility where the installation will take place, such as the customer representative who has been active in the project from the beginning. Be sure the plan identifies who installers should contact to obtain access to restricted areas and locked rooms. Also indicate who can answer questions regarding the installation procedures and network configuration.

Safety Tips

When installing network components, accidents are less likely to happen if you incorporate good safety practices and remind people about them. You should list these safety tips in your installation plan and stress them at your preinstallation meeting:

- Insist that no installers work alone—use the buddy system. If a severe accident occurs, the other person can get help.

- Recommend that installers remove rings and necklaces while installing hardware components. A metal necklace can dangle into a live electrical circuit (or one that is not connected to a power source, but is still energized by charged capacitors) and provide the basis for electrical shock. Rings also conduct electricity or can catch on something and keep you from removing your hand from a computer or component.

- Use proper ladders and safety harnesses if placing antennas on towers or rooftops. There is no reason to take high elevation risks.

- Wear eye protection when using saws or drills.

- Check local OSHA (Occupational Safety and Health Administration) requirements.

Installation Procedures

The plan should clearly describe the procedures for installing components. In some cases, you can simply refer installers to the manufacturer's instructions. Otherwise, write at least the major steps involved in installing each component. You can use the procedures for installing and testing the network outlined in the next sections as a basis.

Tools

Be sure to identify the tools necessary to complete the job. If you have ever constructed a Barbie house, built a patio cover, or worked on a car engine, you certainly realize the need for having the right tools. Not having the proper tools results in time delays looking for the tools or rework needed because you used the wrong tools. The following are tools the installers may require:

○ Wireless installation tools and utilities assist in planning the location of access points and testing wireless connections. The tools are generally available from the applicable wireless product vendor.

○ Two-way radios provide communications among the installation team, especially when spread over a large geographical area.

○ Flashlights are helpful when installing cables above ceilings.

○ Ladders provide access to ceilings and rooftops.

○ Screwdrivers, wire cutters/strippers, crimping tools, and drills help install hardware.

○ Specific test equipment verifies the network installation.

Reference to Design Documentation

The installation will probably require use of design documentation to better understand the overall network configuration. Be sure to indicate the existence of the documentation and how to obtain it.

Schedule

Create a schedule that identifies when to perform each of the installation activities. This helps keep the installation process on schedule. Unfortunately, the best time to install network components is during down-time, such as evening hours and weekends, minimizing disturbances. Hospitals and warehouses never close, but you should plan the installation activities for when the organization is least active.

Resources

Make certain the plan identifies resources needed to perform the installation procedures. Generally, you will not have a staff of technicians with the experience of installing wireless networks. If you plan to perform wireless installations as a service to other companies, then you may want to train existing staff to actually do the implementations. In cases where it is a one-time installation, however, outsourcing the work to a company specializing in network installations is best.

Budget

Create a budget to track expenses related to the installation. The project team may have already prepared a budget during the project planning stages. At this time, the budget may only need to be refined to reflect the installation plan.

Risks

Identify any risks associated with the activities and explain how these risks can be minimized. You might be required, for example, to install 200 wireless LAN connections within a two-day time period. With only two installers, you run the risk of not completing the installation on time. Therefore, you will need to look for additional help to keep on schedule. If someone needs to preapprove your plan, it is best to identify risks and solutions before starting any work.

Coordinating the Installation

Most everyday events require you to coordinate activities. For example, before you, your spouse, your four kids, and pet dog leave on a five-day automobile trip from Dayton to Sacramento to stay a week with relatives, you'd certainly want to communicate with someone at your destination to coordinate items such as sleeping accommodations and eating preferences. Before leaving on the trip, you should let your relatives know that you are on the way. In addition, you would certainly want to hold a predeparture meeting with the family, particularly the kids, and talk about proper behavior while traveling. The installation of a network is probably much easier to pull off than the car trip, but the coordination of activities is similar. As illustrated in figure 10.2, coordination of installation activities include the following:

- **Communicating with the facility's point of contact.** The person designated as the point of contact should understand the installation plan. In fact, he should have been active in developing the schedule to minimize any negative impacts on the organization.

○ **Giving the organization's employees a heads-up.** If you have to install components of the network when the organization's staff is present, announce when, where, and for how long installers will be working within the area. People need time to rearrange their schedules if necessary to accommodate the installation. Have the organization send out a memo or e-mail to announce the installation.

○ **Holding a preinstallation meeting.** The preinstallation meeting gathers together everyone involved in the installation to review procedures. Be sure everyone knows who to contact if problems occur. This meeting is also the best time to remind people about good safety practices.

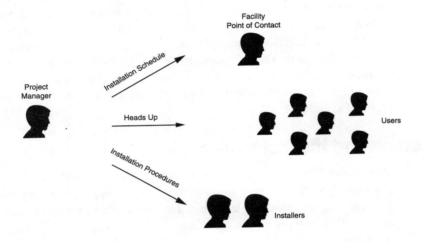

Figure 10.2
Coordinating the installation of the network.

With an installation plan in hand and coordination behind you, the installation may begin.

Installing the Components

When designing the network, as explained in Chapter 8, use a top-down approach. The definition of applications comes first, then network operating system, then medium access, and so on. The installation of the network components, though, should follow a bottoms-up approach. First construct and configure individual components, such as servers, user workstations, and cabling. Then connect the components together, implement communications software, and install and link the applications.

Consider installing the network from the ground up as follows:

1. Install network interface cards in the computers.

2. Install cabling if necessary.

3. Install access points.

4. Establish wide area network connections.

5. Assemble server hardware.

6. Install and configure the network operating system.

7. Install client software on the user workstations.

8. Install applications.

NOTE

Be sure to follow installation procedures supplied by the vendor.

In most cases, the team can install some components in parallel to decrease installation time. For example, installing network interface cards, cabling, and access points at the same time cannot cause any harm. Just be sure to properly test each component, as described later in this chapter, before connecting the pieces together. The point is that you cannot fully access or operate the network operating system and applications until after lower-level components, such as cabling, network installation cards, and access points, have been installed.

If you're installing laser links between buildings, you will probably utilize LCI's LACE or OmniBeam products. The following are tips that LCI offers for installing their point-to-point laser systems:[1]

○ The beam should not be directed near or through electrical power lines or tree branches. Consider tree growth, wind load on trees, and power lines when installing the system. Power lines also sag during warm weather and tighten up during cold weather. Make provisions to discourage nest building in the optical path by birds and insects.

○ Make sure the transmission path is at least ten feet above pedestrian or vehicular traffic—preventing accidental viewing of the laser beam and keeping the signal from being interrupted. Make allowances for any unusual effects that traffic may cause, such as dust clouds.

○ Make sure the transmission path is not shooting through or near exhaust vents, which can cause steam to be blown into the path. This has the same effect as fog on the laser beam.

○ LACE units are fully weatherproofed and are intended to be mounted outdoors. They may be mounted inside buildings, however, and the signal passed through glass windows. When light particles hit a glass surface, some of the light is reflected. With a clear glass window, approximately four percent of the light is reflected per glass surface. If the glass is tinted, the amount of light reflected and absorbed increases. In the case of reflective coatings, the laser light will reflect off of the coating and the light will never be detected at the receiver. Another problem when shooting through glass occurs when it rains. Water droplets on the glass in front of the transmit lens act as additional lenses and can cause the beam to diffuse. Mounting the laser near the top of the window will reduce this problem somewhat, especially if the window has an awning. When passing signals through glass, it is advisable to keep the beam as close to perpendicular to the glass as possible to minimize reflection losses which can reduce signal strength (see fig. 10.3). As the angle of the beam to the glass increases, more and more light is reflected until the critical angle is reached (approximately 42 degrees). Above the critical angle, all the light is absorbed into the glass and no transmission occurs.

○ Avoid East-West orientations. Although LCI uses optical filters in the receiver and has a small angle of acceptance, direct sunlight can overload the units for several minutes a day for a few days per year.

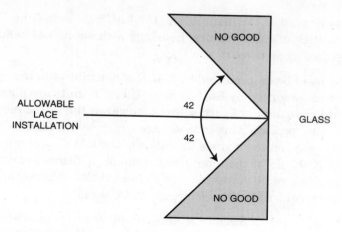

Figure 10.3
The critical angle of a laser beam penetrating glass.

○ Heat from roof tops, air duct vents, air conditioners, glass-faced buildings, and so on can cause a condition known as shimmer. Shimmer will cause the light beam to bend and appear to dance around the receiver. If sufficient heat is present, the beam will deflect enough to miss the receiver altogether, usually for a few milliseconds at a time, and burst error will occur. When mounting on roof-top locations, the preferred location is at the leading edge of the roof with the front of the laser at least six inches over the edge. Placing the laser at this location minimizes the effect of roof heating, heat rising up the side of the building, and snow accumulation in front of the unit. The location also provides access to the rear of the unit for easier set-up and alignment.

○ The movement of LACE units caused by a strong mechanical vibration could cause the system to intermittently go in and out of alignment. Avoid mounting LACE near vibrating machinery such as air conditioning units, compressors, and motors.

○ The LACE units are designed to operate within the temperature range of -10 degrees Fahrenheit to +120 degrees Fahrenheit. Although the units may operate at further extremes, do not do so over an extended period of time. If sustained periods of extraordinarily severe temperatures are normally experienced at the installation, the LACE units may be mounted in environmental housings with additional heating or fan cooling as necessary.

○ The laser beam produced by LACE is not subject to the interference produced by EMI sources. If LACE units are placed within proximity of such sources, however, the LACE electronics may "pick-up" this interference which would then be impressed on the signals to and from the LACE equipment. LACE should be mounted away from large microwave dishes, antennas, radio stations, or any unusual electronic equipment that may be radiating electromagnetic signals.

○ The laser transmission system is employed in point-to-point, line-of-sight applications. To help alleviate the problems with beam wander due to shimmer, the laser beam is purposely diverged by the transmit lens to give a two meter footprint at the receiver. Diverging the laser also helps with maintaining alignment. The received light is focused by a collecting lens onto the photodetector. This lens system has a field of view of only five milliradians, providing some selectivity by cutting down on background light sources. As shown in figure 10.4, the receiver acceptance angle is greater than the transmitter divergence angle, making alignment of the transmitter more critical than alignment of the receiver.

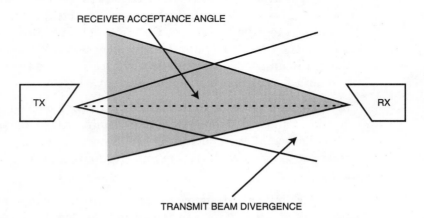

Figure 10.4
Laser system transmitter divergence and receiver acceptance angles.

○ LACE units are designed to project a two meter diameter beam at the receiver. This provides some latitude for beam movement. Unit movement, however, should be kept to an absolute minimum to ensure peak performance. A movement of only 1 mm at the transmitter can divert the beam off of the receiver if the units are installed one kilometer apart.

○ Ideally, LACE units should be mounted on the corner of the building to which it will be attached and preferably to masonry construction. This will provide the most stable arrangement. When transmitting signals over 300 meters, it is not advisable to mount LACE units anywhere except at the corner of the structure. On buildings where a thin metal skin is covering the building, the base for the mounts must be made to the supporting structure or to the metal substructure.

○ Do not mount LACE units on structures which can sway, such as trees, fences, towers, poles, or buildings exceeding 40 stories in height. Always avoid moveable camera mounts.

○ Do not mount LACE units to wooden structures. The expansion/contraction properties of these materials through precipitation and temperature make them good sources for movement and should be avoided. For example, high humidity will cause the units to go out of alignment due to the wood expanding.

○ Make sure that when LACE is mounted there are no ledges in front of the laser that might be used by roosting birds. Ledges can also cause a problem in rain or snow. Water bouncing up from the ledge onto the optics or snow buildup in front of the optics will diminish performance.

○ Once a stable mounting position has been chosen to mount the laser, the actual design of the mount needs to be considered. The mount should always provide a flat surface with contact on the entire bottom surface of the LACE equipment. LACE units should be mounted to supports with a 6"× 6"× 3/8" metal plate welded to brackets which attach to the supporting structure. These mounts must be substantial enough to resist any movement. Typical light-duty CCTV camera mounts are not acceptable. To test the mount and support structure for a LACE system, push on the mount and the LACE unit with about five to ten pounds of force. This should not disrupt the alignment. If, for example, a person on a ladder leaning against a wall supporting LACE is enough to disrupt alignment, the structure is not strong enough to provide a good, long-term mount.

Testing the Installation

In the software world, testing is extremely important—countless problems have resulted from defects in software. You might have heard, for example, about the farmer in a remote area of Nebraska who one day received thousands of the same copy of Time magazine? Apparently, the publisher's label printing software had a defect that made it spend hours printing the same address. Defects in networks might not cause similar incidents, but improper configurations and unforeseen propagation impairments can easily discontinue the network's operation. This is especially serious in hospitals where a doctor's or rescue person's access to information can create a life-or-death situation.

Concepts of Testing

Several definitions for testing exist. Some people say that testing is for checking if the system offers proper functionality, and others say the reason for testing is to find conditions that make the system fail. Actually, testing is a combination of the two: it ensures the network behaves as expected and that no serious defects exist. Figure 10.5 illustrates the elements of network testing.

Test Cases

A test case represents an action you perform and its expected result. For example, one test case might determine whether access to a database meets performance requirements. The action would be to run a particular query, and the expected result would be the time it takes for the query to return the corresponding data.

How do you write test cases? First, be sure to review the later sections on performing unit, integration, system, and acceptance testing before writing test cases. Then, referring to the network requirements and design defined earlier in the project, describe the tests necessary to ensure a network that behaves adequately. The following are attributes of a good test case:

○ Has a good chance of uncovering a defect

○ Can be performed easily

○ The expected result is verifiable

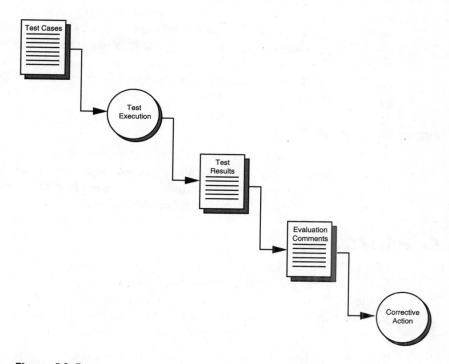

Figure 10.5
The elements of network testing.

Test Execution

With a complete set of test cases, you're ready to run the tests. As you'll see later, testing takes place throughout the installation phase. You might have noticed that building contractors must carefully inspect the foundation before building the structure on top; otherwise, the building itself could hide support structure defects that could later cause a disaster. Network testing is similar;

you should fully test the radio connectivity and cabling before installing and testing their interaction with the network operating system and application.

Test Results

The outcome of performing test cases is test results. Of course, you hope everything checks out okay. Poor results indicate the need for rework, meaning a design modification is necessary and components need to be reinstalled or reconfigured. This will take time to complete, possibly extending the project. Recording test results for observation later when supporting the network is a good idea. Test results offer baseline measurements support staff can use to aid in troubleshooting future problems.

Evaluation Comments

After obtaining the test results related to the portion of the network under testing, compare them to the expected values identified in the test case. The evaluation comments should explain any differences and, if necessary, recommend corrective action.

Corrective Actions

Figure 10.6 shows an example of a form you can use to document test activities. The data provides baseline information for later testing. When problems arise during the network's operational life, you can run tests again and compare them to the ones run during installation, which can also help pinpoint problems.

The best method of testing the network installation is to follow a bottoms-up approach by performing the following types of tests:

1. Unit testing
2. Integration testing
3. System testing
4. Acceptance testing

Test Case				
Test Description	Expected Result	Actual Result	Eval. Comments	Corrective Action

Figure 10.6
A sample form for documenting test activities.

Performing Unit Testing

Unit testing verifies the proper internal operation or configuration of individual network components, such as network interface cards, servers, cables, printers, and so forth. You should unit test each component—before trying to make them work with other parts of the system—to ensure the components themselves are not defective. Knowing that individual parts work to make future troubleshooting problems easier. Figure 10.7 illustrates the concept of unit testing.

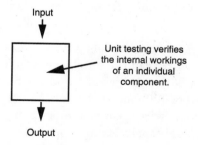

Figure 10.7
Unit testing makes future troubleshooting easier.

Ideally, you would want to fully test all possible functions and configurations of each unit; however, that's usually not feasible. The following sections offer examples of unit tests you should perform.

Testing Individual Components

Be sure to test the operation of each component, such as printers, servers, and access points, before integrating them with other components. Most components have built-in self-tests that run whenever you turn the device on, or they have test utilities that you can run manually. Therefore, you usually won't need to develop specialized test cases for most individual units. Proxim's RangeLAN2, for example, comes with a utility called RL2SETUP which verifies whether you've chosen an I/O address, IRQ, or memory window that may conflict with other hardware.

Testing Category 5 Cable Installations

Cable problems rank high as causes of networking troubles. Mechanical elements, such as cabling, connectors, and wall plates, tend to fail more often than active electronic devices such as network adapters and switches. Approximately 85 percent of cable problems arise from the installation; therefore, be sure to fully test cable installations. Cable faults result from improper splices, improper connector attachments, lack of termination, and corrosion.

The good news is cable problems are relatively easy to find, especially if you utilize an effective cable tester conforming to TIA's Technical Service Bulletin (TSB) 67, published by the Link Performance Task Group of the Telecommunications Industry Association (TIA). This TSB is not a standard, but it does describe how to test Category 5 twisted-pair cable. You should definitely consider TSB-67 when selecting a cable tester.

TSB 67 addresses two link configuration models: Channel Link and Basic Link. The *Channel Link* consists of the patch cords that connect the access points to the horizontal wiring, and the horizontal wiring itself can span a total of 100 meters. Channel Link testing covers a range that verifies wiring connections up to the user's interface. The *Basic Link* only includes the horizontal wiring and two 2-meter tester equipment cords and can be 90 meters long.

Installation crews commonly perform Basic Link testing after laying the cabling.

The authors of TSB 67 chose two levels of accuracy for testing links: Level I for low accuracy and Level II for high accuracy. These two accuracy levels take into consideration the test configurations you implement for testing the Basic and Channel Links. For instance, Channel Link testing almost always requires the use of an RJ45 interface attached directly to your tester. The problem is the RJ45 interface offers unpredictable crosstalk and affects the accuracy of crosstalk measurements. This type of test, therefore, would only need Level I testing. On the other hand, Basic Link testing enables you to interface the tester to the cable via a connector having much lower crosstalk, such as a DB-9 or DB-25 connector. Thus, with Basic Link testing, it is possible to run the more accurate Level II tests.

After installing Category 5 cabling, test the installation by performing the following tests that TSB-67 recommends.

Wiremap

The wiremap test ensures a link has proper connectivity by testing for continuity and other installation mistakes, such as the connection of wires to the wrong connector pin. For example, if you don't wire an RJ45 connector exactly according to a standard, such as EIA/TIA 568A's T568A or T568B wiring scheme, then you might produce split pairs. A *split pair* occurs when you attach the connector in a way that a wire pair consists of one lead from one twisted pair and another lead from a different twisted pair, creating a pair of wires that are untwisted. The split pair might result in an excessive amount of external noise interference and crosstalk, which will cause transmission errors. Most cable testers perform wiremap tests to detect this type of cable problem.

Link Length

Link length measurements identify whether a cable meets the length limitations. Cable testers utilize a Time-Division Reflectometer (TDR), which measures the length of a cable. The operation of a TDR is shown in figure 10.8. The TDR emits a pulse at one end of the cable, which travels to the opposite end of the cable, and then reflects back to the TDR. The TDR measures the propagation time and calculates the cable length based on an average wave propagation rate.

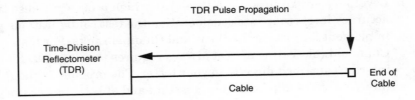

Figure 10.8
Operation of a time-division reflectometer (TDR) test.

Several products on the market run TDR tests on metallic or optical-fiber cable. Tektronix TS100 Option 01 Metallic TDR, for example, tests LocalTalk, Type 1 and 3, Category 3, 4, and 5, and thin and thick coax cables. This test set finds shorts, opens, and breaks in the cable. The Tektronix TFP2A Fibermaster OTDR tests single-mode and multimode fiber-optic cables.

Attenuation

Attenuation tests ensure the cabling will offer acceptable attenuation over the entire operating frequency range. If too much attenuation is present, digital signals sent throughout the cable will experience rounding, resulting in transmission errors. Cable testers examine attenuation by measuring the effects of sending a series of signals that step through the cable's operating frequency bandwidth. For Category 5 testing, most cable testers cover bandwidth of 1 MHz–100 MHz by taking readings in 1 MHz increments, certifying whether the cable meets specifications in the part of the frequency spectrum where the signal mostly resides. The Microtest Pentascanner is an example of a cable tester that measures attenuation on Category 3, 4, and 5 cable.

Near-End Crosstalk (NEXT)

Crosstalk is the crossing of current from one wire to a nearby wire, causing transmission errors. Near-end Crosstalk (NEXT) is a specific case where signals at one end of the link interfere with weaker signals coming back from the recipient. The amount of

NEXT varies erratically as you sweep through the operating bandwidth of a cable. For an accurate measurement, cable testers record NEXT by stepping though the cable's operating frequency range at very small increments. For Category 5 cable, TSB-67 recommends a maximum step size of 0.15 MHz for lower frequencies and 0.25 MHz for higher frequencies within the 1–100 MHz frequency range. This requires a fast instrument to take the hundreds of samples necessary. Fluke's DSP-100 handheld cable tester is an example of an incredibly fast NEXT tester. The DSP-100 utilizes digital signal processing to increase its speed and allow samples to be taken at close 100 KHz intervals. The DSP-100 performs all tests required by TSB-67 for a 4-pair cable in under 20 seconds. The DSP-100 not only identifies the presence of crosstalk, but also locates its source.

If any defects are found through unit testing, correct the problems of each unit before integrating them together with other components.

Performing Integration Testing

The concept of integration testing, as shown in figure 10.9, is to test the network as you connect the components together. In this figure, integration testing would verify that components A and B work together okay. Then, after component C is installed, integration testing would verify that all three components work together acceptably.

Unit testing guarantees the proper functioning of individual components, but the testing is not enough to certify a network. Integration testing goes a step further and ensures these components operate together. In other words, integration testing spans multiple units as shown in figure 10.9. For example, you may have unit tested a PC containing a wireless network interface card, an associated access point, and a server located on the wired network and found all of them operating sufficiently. However, performing an integration test, such as attempting to log into the server from the PC via the access point, may fail. The reason could be because the network interface card and access point were set to different channels, not allowing a connection.

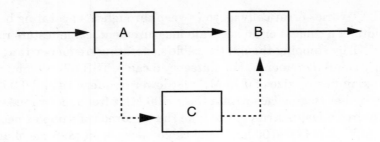

Figure 10.9
The concept of integration testing.

> **NOTE**
>
> Performing integration testing constantly as you add components to the network is important. You can then find problems soon before they become buried. For instance, the previous problem is fairly easy to find because there were a limited number of components involved. If you were to finish the entire implementation first and then run tests, finding the problems would be more difficult. If the final configuration of the preceding example includes TCP/IP access to an application residing on the server via a WAN as well, troubleshooting would be much more complex because of the additional components. Thus, be sure to include integration testing *as* you build the network.

As with unit testing, the ideal is to verify all possible functions across the set of components you're testing; however, that's not feasible in most cases. To help you narrow possibilities to a workable set, the following are some examples of integration tests you should consider performing:

○ Capability to roam from one radio cell to another

○ Capability to roam throughout the designated coverage area

○ Capability of a remote host at the end of a TCP/IP connection to respond to a continuity test, such as a Ping

As with unit testing, correct any defective installations before pressing on with further integration or system testing.

Performing System Testing

System testing determines whether the completed network installation is capable of satisfying the requirements specified at the beginning of the project. Thus, system testing requires you to first install all components (and perform appropriate unit and integration tests). Figure 10.10 illustrates the concept of system testing, which is the final testing done before handing the system over to the users for acceptance. For example, after installing all network components (the clients' terminals, access points, ethernet networks, WAN, servers, and applications), you should check whether the software on the client workstation communicates correctly with the application on the server.

Figure 10.10
The concept of system testing.

The scope of system testing depends on the potential for disaster if the network fails. As an example, if you're building a network to support an information system onboard a manned spacecraft destined for Mars, then you would want to perform exhaustive testing to discover and fix any defects so that unfixable problems don't arise en route. The main reason for the thoroughness is because human lives are at stake if the system fails. Most Earth-based systems without human lives on the line will not require this extreme testing. Be sure, however, to develop test cases that exercise the system from one extreme to another. The goal is to develop and execute system tests that verify, at the minimum, the following system attributes:

○ Capability of users to access appropriate applications from terminals and PCs

    ○ Capability to support all security requirements

    ○ Capability to meet performance requirements

    ○ Capability to interface with all external systems

If the testing of these attributes provides unfavorable results, take corrective actions and then retest the portions of the system that required modification.

Performing Acceptance Testing

After the project team fully tests the system, it's time for the customer to perform acceptance testing, which involves actual users running tests to determine whether the implementation is acceptable. These tests should focus on verifying whether the system functions as specified in the requirements. This does not require the same level of detail as system tests do. Acceptance testing is done at a much higher level to ensure users can run the appropriate applications while performing their jobs. For example, Mary will be using a wireless terminal tied to a central database to access inventory records as she stocks the shelves. Acceptance testing will determine whether she can actually enter and retrieve data from the database while handling the stock.

Deploying the network/system to the entire population until a cross section of users performs the acceptance tests is not best. Most people refer to this as a *system pilot*. Acceptance testing as a pilot of the implementation is advisable if any of the following conditions are true:

    ○ The implementation spans multiple geographic locations.

    ○ The network supports mission-critical applications.

In these cases, great risks in loosing productivity and valuable information exist if defects in the system occur. For instance, if you deploy a wireless inventory system at six warehouses and find that other existing devices interfere, then all warehouses stand to lose some productivity until the problem can be resolved. A pilot system at one of the warehouses would have identified the problem, and you could have fixed it before deploying the system to the remaining sites. The drawback of pilot testing, though, is that it delays the deployment of the system to the users not participating in the pilot. This could make it impossible to meet schedule deadlines. However, if the conditions above exist, running a pilot test will be worth the wait.

Finalizing the Project

After the user organization accepts the system, the project might seem to be over. Right? Actually, no. There are still some tasks left, including the following:

○ Updating documentation

○ Training the users

○ Transferring the system to operational support

○ Evaluating the outcome of the project

Updating Documentation

During the installation and testing phase, the team may have made changes to the design or layout of the network as a result of corrective actions to failed tests. Therefore, the team might need to update documentation, such as design specifications. In some extreme cases, requirements might need updating if the installed system can't support desired requirements as expected. Most companies refer to these updated documents as "installed" or "red-lined" drawings. These drawings provide an accurate set of documentation for support staff to use when troubleshooting or modifying the system.

Training Users

As described in Chapter 9, "Preparing for the Support of a Wireless Network," training users and, if necessary, the support staff, is extremely important. Training strengthens the interface between the system and the users. If possible, offer the training before or during the system installation. Training prepares specific users for performing the acceptance testing and ensures all users are ready to start using the system when it's operational. You can implement a small implementation of the system in a classroom and teach people how to use the system before it is actually deployed to the rest of the company.

Transferring the Network to Operational Support

While implementing the network, the project team provides support for the network, such as the creation of user accounts, and trouble-shooting and repair actions. In fact, during the requirements phase and acceptance testing, potential users generally assume that members of the project team will always be providing support. In some cases, this might be true. But, regardless of whether the team will be providing operational support, be sure to clearly transfer support of the operational network from the project team to appli-cable people and organizations. This transfer clearly marks the end of the project and ensures users having problems with the system will call upon the right people for assistance. Be certain as part of the transfer that the operational support staff have copies of network documentation, such as designs and support plans. Figure 10.11 illustrates the concept of transferring the network to opera-tional support.

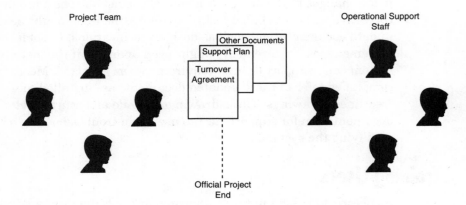

Figure 10.11
Transferring the network to operational support.

This transfer should mark the completion of a successful wireless network implementation. Soon, the numerous benefits of wireless networking will be apparent and your efforts will be justified.

Evaluating the Outcome of the Project

○ Did upper management continue to support the project to the end?

○ Did the requirements phase go smoothly?

○ Did all members of the team communicate effectively within the team and with other individuals and groups?

○ Were there any problems associated with the mechanics of product procurement?

○ Was operational support in place before users began using the system?

○ Did the training properly prepare users to operate the system?

○ Was the project completed on schedule and within budget?

○ Do the users feel the system will enhance their performance?

○ Does the implemented system perform as expected by the project team and the users?

---

[1] These tips are reprinted by permission from LCI, Inc.

PART IV

Appendixes

A. Wireless Networking Vendors and Service Companies

B. Wireless Networking Products and Services

C. Standards Organizations and Industry Groups

D. Wireless Networking in Healthcare

E. Glossary

Wireless Networking Vendors and Service Companies

This appendix identifies companies that sell network components and services that might assist you in the implementation of a wireless network.

Network Product and Service Companies

The following companies specialize in wireless networking products and services:

AeroComm Wireless
13228 W. 99th St.
Lenexa, KS 66215
Tel: 800-492-2320, 913-492-2320
FAX: 913-492-1243
Wireless LAN Products (GoPrint)

Aerotron-Repco Sales, Inc.
2400 Sand Lake Rd.
Orlando, FL 32809-7666
Tel: 800-950-5633, 407-856-1953
FAX: 407-856-1960
Wireless WAN and MAN Products

Aironet Wireless Communications, Inc.
367 Ghent Rd., Suite 300
PO Box 5292
Fairlawn, OH 44334-0292
Tel: 800-3-WIRELESS, 216-665-7900
FAX: 216-665-7922
Internet: sales@aironet.com
Wireless LAN Products (ARLAN)

Alps Electric (USA), Inc.
3553 N. First St.
San Jose, CA 95134
Tel: 800-825-2577, 408-432-6000
FAX: 408-432-6035
Wireless LAN Products

Ameritech Cellular Services
2000 W. Ameritech Center Dr.
Hoffman Estates, IL 60195-5000
Tel: 800-MOBILE-1, 708-706-7600
Wireless Data Questions: 800-669-4730
FAX: 708-765-3702
Wireless WAN Products and Services

ARDIS
300 Knightsbridge Pkwy.
Lincolnshire IL 60069
Tel: 708-913-1215
FAX: 708-913-1453
Wireless WAN Services

Astronet Corporation
37 Skyline Dr.
Lake Mary, FL 32746-6214
Tel: 407-333-4900
FAX: 407-333-4966

AT&T Global Business Communications Systems
211 Mt. Airy Rd.
Basking Ridge, NJ 07920
Tel: 800-325-7466, ext. 837
FAX: 908-953-4385
WWW: http://www.att.com/gbcs

AT&T Global Information Solutions
1700 S. Patterson Blvd., PCD/6
Dayton, OH 45479-0001
Tel: 513-445-7197
FAX: 513-445-2468
Wireless LAN Products

AT&T Wireless Services
Messaging Division
110 110th Ave. NE, Suite 200
Bellevue, WA 98004
Tel: 206-990-4481
FAX: 206-990-4299
WWW: http://www.mccaw.com
Wireless WAN Services

AT&T Wireless Services
Wireless Data Division
10230 NE Points Dr.
Kirkland, WA 98004
Tel: 800-IMAGINE
FAX: 206-803-4601
WWW: http://www.airdata.com
Wireless WAN Services

A.T. Schindler Communications, Inc.
101-21 Antares Dr.
Ottawa, Canada K2E 7T8
Tel: 613-723-1103
FAX: 613-723-6895
WWW: http://www.firlan.com

BASS, Inc.
2211 Arbor Blvd.
Dayton, OH 45439
Tel: 800-533-2277
FAX: 513-293-9756
Wireless LAN Products

Breeze Wireless Communications, Inc.
[BreezeCOM (formerly Lannair, Ltd.)]
2195 Faraday Ave., Suite A
Carlsbad, CA 92008
Tel: 619-431-9880
FAX: 619-431-2595
Internet: n.america-sales@breezecom.com
WWW: http://www.breezecom.com
Wireless WAN Products
California Microwave

Cylink
910 Hermosa Ct.
Sunnyvale, CA 94086
Tel: 800-533-3958, 408-735-5800
FAX: 408-735-6643, FAX on demand: 408-735-6614
Internet: info@cylink.com
WWW: http://www.cylink.com
Wireless LAN and MAN Products (AirLink)

Data Broadcasting Corporation
1900 S. Norfolk St., Suite 150
San Mateo, CA 94403
Tel: 415-571-1800
FAX: 415-571-8507
WWW: http://www.dbc.com

Data Critical Corporation
2733 152nd Ave. NE, Bldg. 5F
Redmond, WA 98052
Tel: 206-885-3500
FAX: 206-885-3377
WWW: http://www.datacrit.com

Dataradio
DATARADIO is a registered trademark of DATARADIO, INC.
6160 Peachtree Dunwoody Rd., Suite C-200
Atlanta, GA 30328
Tel: 404-392-0002
FAX: 404-392-9199
Internet: info@dataradio.com

Digital Ocean, Inc.
11206 Thompson Ave
Lenexa, KS 66219
Tel: 913-888-3380
FAX: 913-888-3342
Internet: Support@DigOcean.com
WWW: http://www.digocan.com
Wireless WAN Products

Digital Wireless Corporation
One Meca Way
Norcross, GA 30093
Tel: 770-564-5540
FAX: 770-564-5541
Internet: mkting@digiwrls.com
Wireless LAN Products

E.F. Johnson
Corporate Headquarters
11095 Viking Drive
Minneapolis, MN 55344-7292
Tel: 800-328-3911 ext. 380

E.F. Johnson Canada, Inc.
633 Granite Court
Pickering, Ontario L1W 3K1
Tel: 800-263-4634
Wireless MAN Products

Eiger Labs
1237 Midas Way
Sunnyvale, Ca. 94086
Tel: 408-774-3456
Wireless LAN Products

Epson America, Inc.
OEM Division
20770 Madrona Ave.
Torrance, CA 90503
Tel: 310-787-6300
FAX: 310-787-5350
WWW: http://www.epson.com

Ericsson
1 Triangle Dr.
PO Box 13969
Research Triangle Park, NC 27709
Tel: 919-990-7000
FAX: 919-990-7444
WWW: http://www.ericsson.com

Ericsson GE Mobile Communications, Inc., a division of Telefon AB LMEricsson Wireless Computing
15 E. Midland Ave.
Paramus, NJ 07652
Tel: 800-223-6336, 201-890-3600, 201-265-6600
FAX: 201-265-9115

ETE, Inc.
12526 High Bluff Dr., Suite 300
San Diego, CA 92130
Tel: 619-793-5400
FAX: 619-793-6060
WWW: http://www.ete.com

Extended Systems, Inc.
5777 N. Meeker Ave.
Boise, ID 83713
Tel: 800-235-7576, 208-322-7575, 406-587-7575
FAX: 208-377-1906
Wireless LAN Products

FreeWave Technologies, Inc.
1880 Flatiron Court
Boulder, CO 80301
Tel: 303-444-3862
FAX: 303-786-9948

GRE America, Inc.
25 Harbor Blvd.
Belmont, CA 94002
Tel: 800-233-5973, 415-591-1400
FAX: 415-591-2001

GRE Japan, Inc.
6-2-15 Roppongi, Minato-Ku
Tokyo, JAPAN 106
Tel: 03-3404-3636
FAX: 03-3405-5387
Wireless LAN Products

GTE Mobilenet
245 Perimeter Center Pkwy.
Atlanta, GA 30346
Tel: 770-391-8000
FAX: 770-395-8733
Wireless WAN Products and Services

IBM Wireless
700 Park Office Rd., Hwy. 54
Building 662
Research Triangle Park, NC 27709
Tel: 919-543-7708
FAX: 919-543-5568
Wireless LAN Products

Inficom, Inc.
645 Southcenter, Suite 343
Seattle, WA 98188-2836
Tel: 206-865-9753
FAX: 206-562-6066
Internet: inficom@inficom.com
WWW: http://www.inficom.com

InfraLAN Wireless Communications
18 Kinsley Rd.
Acton, MA 01720
Tel: 508-266-1500
FAX: 508-263-1011
Wireless LAN Products

Infralink of America, Inc.
1925 N. Lynn St., Suite 703
Arlington, VA 22209
Tel: 404-409-7858
FAX: 404-449-9236

Intel Corporation
2200 Mission College Boulevard
Santa Clara, CA 95052-8119
Tel: 800-538-3373

Intermec Corporation
6001 36th Ave. SE
Everett, WA 98203
Tel: 800-347-2636
FAX: 206-348-2833
WWW: http://www.intermec.com

Itronix Corporation
S. 801 Stevens St.
Spokane, WA 99204
Tel: 800-555-4104
FAX: 216-873-2099
Internet: http://www.itronix.com

JTECH, Inc.
6413 Congress Ave.
Boca Raton, FL 33487
Tel: 800-321-6221
FAX: 407-997-5753
WWW: http://www.jtechinc.com

K&M Electronics, Inc.
11 Interstate Drive
West Springfield, MA. 01089
Tel: 413-781-1350
FAX: 413-737-0608
Wireless LAN Products

Laser Communications, Inc.
1848 Charter Lane, Suite. F
PO Box 10066
Lancaster, PA 17605-0066
Tel: 800-527-3740, 717-394-8634
FAX: 717-396-9831
Internet: lasercom@epix.net
WWW: http://www.lasercomm.com/lasercomm/
Wireless MAN Products

Logistics Systems Engineering
PO Box 6299
Annapolis, MD 21401
Tel: 410-268-9101
FAX: 410-268-9107

MCOM Network Systems
1748 Independence Blvd E-1
Sarasota, FL 34234-2152
Tel: 813-358-9283 (wave)
For Faxback, press 4
Wireless MAN Products

Medical Communications System
20 E. Springfield St.
Boston, MA 02118
Tel: 617-465-6500
FAX: 617-424-9189
Wireless Healthcare Systems

Meteor Communications Corporation
8631 S. 212th St.
Kent, WA 98031
Tel: 206-872-2521
FAX: 206-872-7662
Wireless WAN Products

Metricom, Inc.
980 University Ave.
Los Gatos, CA 95030-2375
Tel: 800-434-4460, 408-399-8200
FAX: 408-399-8274
Internet: info@ricochet.net
WWW: www.ricochet.net
Wireless WAN Products and Services (Ricochet)

Microwave Data Systems
175 Science Pkwy.
Rochester, NY 14620
Tel: (716) 442-4000
Wireless MAN Products

Mobile Satellite Products Corporation
55 Commerce Dr.
Hauppauge, NY 11788-3931
Tel: 516-273-4455
FAX: 516-273-4583
Internet: Info@MobileSat.com
WWW: http://www.mobilesat.com
Satellite Products

Monarch Marking
170 Monarch Lane
Miamisburg, Ohio 45342
Tel: 800-543-6650
FAX: 513-865-6605
Internet: Info@mmsday.com
Portable Data Collection and Scanning Products

Motorola—Wireless Data Group
50 E. Commerce Drive, Suite M-4
Schaumburg, IL 60173
Tel: 800-WIRELESS-8, 708-576-8213
FAX: 708-576-8940
800-624-8999 ext. 105
WWW: http://www.mot.com/wdg/
Wireless LAN and MAN Products

Multipoint Networks
19 Davis Ave.
Belmont, CA 94002-3001
Tel: 415-595-3300
FAX: 415-595-2417
Wireless MAN Products

NEC America, Inc.
Satellite Communications Products
14040 Park Center Rd.
Herndon, VA 22071
Tel: 703-834-4150
FAX: 703-834-4757
WWW: http://www.nec.com

Nettech Systems, Inc.
58 Wall St.
Princeton, NJ 08540
Tel: 609-683-0100
FAX: 609-683-5019
Wireless Applications

Norand Corporation
550 Second St. SE
Cedar Rapids, IA 52401
Tel: 800-553-5971, 319-369-3100
FAX: 319-369-3453

NORCOM Networks Corporation
3650 131st Ave. SE, Suite 510
Bellevue, WA 98006
Tel: 800-676-9951
FAX: 206-649-9236
WWW: http://www.norcom.net

Northern Telecom (Nortel)
PO Box 833858
2221 Lakeside Blvd.
Richardson, TX 75083-3858
Tel: 800-4-NORTEL
WWW: http://www.nt.com

Notable Technologies, Inc.
411 108th Ave. NE, Suite 1000
Bellevue, WA 98004
Tel: 206-455-4040
FAX: 206-445-4440
WWW: http://www.airnote.net

Nova CounterElectronics, Inc.
15201 N. 1st. St.
Raymond, NE 68428
Tel: 402-785-2838
FAX: 402-785-3001
Internet: sales@novacorp.com
WWW: http://www.novacorp.com
Wireless LAN Products

O'Neill Connectivities, Inc.
607 Horsham Rd.
Horsham, PA 19044
Tel: 800-624-5296, 215-957-5408
FAX: 215-957-6633
Wireless LAN Products

On Target Mapping
1051 Brinton Rd.
Pittsburgh, PA 15221
Tel: 800-700-6277
FAX: 412-241-7709

ORBCOMM
21700 Atlantic Blvd.
Dulles, VA 20166
Tel: 800-ORBCOMM
FAX: 703-406-3504
WWW: http://www.orbcomm.net

PageMart, Inc.
6688 N. Central Exp. #800
Dallas, TX 75206
Tel: 800-593-4953, 214-750-5809
WWW: http://www.pagemart.com

PageSat, Inc.
992 San Antonio Rd.
Palo Alto, CA 94303
Tel: 800-227-6288, 415-424-0384
Satellite Products

Palmtree Products, Inc.
145 Washington St.
Norwell, MA 02061
Tel: 617-871-7050
FAX: 617-871-6018
GPS Products

PCSI
9645 Scranton Rd.
San Diego, CA 92121
Tel: 800-933-PCSI, 619-535-9500
FAX: 619-535-9235
WWW: http://www.pcsi.com

Persoft, Inc.
465 Science Drive
Madison, WI 53711
Tel: 800-368-5383, 608-273-6000
FAX: 608-273-8227
Wireless MAN Products

Photonics Corporation
2940 N. First St.
San Jose, CA 95134
Tel: 800-628-3033, 408-955-7930
FAX: 408-955-7950
Wireless LAN Products

Precision Tracking FM, Inc.
15001 E. Beltwood Pkwy.
Dallas, TX 75244
Tel: 800-880-3335, 214-991-1772
FAX: 214-991-1752
WWW: http://www.ptfm.com

Priority Call Management
226 Lowell St.
Wilmington, MA 01887
Tel: 508-658-4400
FAX: 508-658-3809
Wireless Applications

Proxim, Inc.
295 N. Bernardo Ave.
Mountain View, CA 94043
Tel: 800-229-1630, 415-960-1630
FAX: 415-960-1984
Wireless LAN and MAN Products (RangeLAN and RangeLINK)

Psion, Inc.
150 Baker Ave.
Concord, MA 01742
Tel: 508-371-0310
FAX: 508-371-9611
WWW: http://www.psioninc.com

Qualcomm, Inc.
10555 Sorrento Valley Rd.
San Diego, CA 92121
Tel: 619-587-1121
Internet: info@qualcomm.com
WWW: ftp.qualcomm.com

RACOTEK
7301 Ohms Lane, Suite 200
Minneapolis, MN 55439
Tel: 612-832-9800
FAX: 612-832-9383

RadioMail Corporation
2600 Campus Dr.
San Mateo, CA 94403
Tel: 415-286-7800
FAX: 415-286-7805
WWW: http://www.radiomail.net
Wireless WAN Applications

RAM Mobile Data
10 Woodbridge Ctr. Drive, Suite 950
Woodbridge, NJ 07095
Tel: 908-602-5500 (800-726-3210)
FAX: 908-602-5242 (800-763-1110)
Internet: RAM.Info@ram.com
WWW: http://www.ram-wireless.com
Wireless WAN Services

Rockwell
400 Collins Rd. NE
Cedar Rapids, IA 52498-0001
Tel: 800-288-3150, 319-395-8113
FAX: 319-395-1766
WWW: http://www.cca.rockwell.com

Seaboard Electronics
70 Church St.
New Rochelle, NY 10805-3204
Tel: 914-235-8073
FAX: 914-235-8369

Siemens Stromberg-Carlson Information Services
900 Broken Sound Pkwy NW
Boca Raton, FL 33487
Tel: 407-955-5000
FAX: 407-955-8351
Internet: incoming@ssc.siemens.com
WWW: http://www.ssc.siemens.com

Sierra Wireless
#260-13151 Vanier Place
Richmond, BC V6V 2J2
Tel: 604-231-1109
FAX: 604-231-1109
WWW: http://www.sierrawireless.com

SilCom Technology
5620 Timberlea Blvd.
Mississauga, Ontario L4W 4M6
Tel: 905-238-8822
Wireless LAN Products

SkyTel
1350 I St. NW, Suite 1100
Washington, DC 20005
Tel: 800-SKY-USER, 800-759-8737, 800-456-3333, 202-408-7444
Pager Products

Socket Communications
6500 Kaiser Drive
Fremont, CA. 94555
Tel: 510-744-2700
FAX: 510-744-2727

Solectek Corporation
6370 Nancy Ridge Drive, 109
San Diego, CA 92121
Tel: 800-437-1518, 619-450-1220
FAX: 619-457-2681
Wireless LAN Products

Southwest Microwave, Inc.
2922 S. Roosevelt St.
Tempe, AZ 85282-2042
Tel: 602-968-5995
FAX: 602-894-1731
Wireless MAN Products

Spectralink Corporation
1650 38th St. 202E
Boulder, CO 80301
Tel: 303-440-5330
FAX: 303-447-2013

Spectrix Corporation
106 Wilmot Rd., Suite 250
Deerfield, IL 60015-5150
Tel: 800-710-1805, 847-317-1770
FAX: 847-317-1517
Internet: sales@spectrixcorp.com
WWW: http://www.spectrixcorp.com
Wireless LAN Products

Steinbrecher Corporation
30 North Ave.
Burlington, MA 01803-3398
Tel: 617-273-1400
FAX: 617-273-4160

Symbol Technologies, Inc.
One Symbol Plaza
Holtsville, NY 11742-1300
Tel: 800-SCAN 234, 516-738-2400
FAX: 516-738-2831
WWW: http://www.symbol.com:80
Wireless LAN Products

TAL (Tetherless Access Ltd.)
930 E. Arques Ave.
Sunnyvale, CA 94086-4552
Tel: 408-523-8000
FAX: 408-523-8001
Email: info@tetherless.com
WWW: http://tetherless.net/
Wireless Services

TASC, Inc.
2555 University Drive
Fairborn, OH 45324
Tel: 513-426-1040
FAX: 617-426-8888
WWW: http://tasc.com
Wireless Network Consulting

Teledesign Systems, Inc.

1710 Zanker Rd.
San Jose, CA 95112-4215
Tel: 408-436-1024
FAX: 408-436-0321

Tellabs

4951 Indiana Ave.
Lisle, IL 60532
Tel: 708-969-8800
FAX: 708-852-7346
WWW: http://www.tellabs.com

Telxon Corporation

3330 W. Market St.
PO Box 5582
Akron, OH 44334-0582
Tel: 800-800-8008, 216-867-3700
FAX: 216-873-2099
Internet: sales@telxon.com
WWW: http://www.telxon.com
Wireless LAN Products

Toko America, Inc.

1250 Feehanville Dr.
Mt. Prospect, IL 60056
Tel: 800-PIK-TOKO
FAX: 708-699-1194

U.S. Paging Corporation

1680 Route 23 N.
Wayne, NJ 07470
Tel: 800-473-0845, 201-305-6000, 201-305-1462
Paging Services

WaveLAN (Lucent Technologies)

111 Madison Ave.
Morristown, NJ 07960
Tel: 800-ATT-WAVE
WWW: http://www.wavelan.com
Wireless LAN and MAN Products

WDC (Wireless Data Corporation)
5380 Peachtree Industrial Blvd., Suite 247
Norcross, GA 30071
Tel: 770-447-4990
FAX: 770-447-1680
WWW: http://www.wireless-data.com

Windata Corporation
543 Great Rd.
Littleton, MA 01460-1208
Tel: 508-952-0170
FAX: 508-952-0168
WWW: http://www.best.com
Wireless LAN and MAN Products

Wireless Connect
2177 Augusta Pl.
Santa Clara, CA 95051
Tel: 408-296-1546
FAX: 408-296-1547
WWW: http://www.wirelessip.net/

Xetron Corporation
460 W. Crescentville Rd.
Cincinnati, OH 45246
Tel: 513-881-3100
FAX: 513-881-3379
WWW: http://www.xetron.com

XIRCOM
2300 Corporate Center Dr.
Thousand Oaks, CA 91320-1420
Sales: (800) 438-4526
Tel: 805-376-9300
FAX: 805-376-9311
Wireless LAN Products

Zenith Data Systems
2150 E. Lake Cook Rd.
Buffalo Grove, IL 60089
Tel: 708-808-5000
FAX: 708-808-4434
WWW: http://www.zds.com
Wireless LAN Products

Other Network Product Companies

The following companies specialize in wired network components:

3Com Corporation
PO Box 58145, 5400 Bayfront Plaza
Santa Clara, CA 95052-8145
Tel: 800-638-3266, 408-764-5000
FAX: 408-764-5001
Product Types:
>   Bridges/Routers/Gateways
>   Hubs
>   Repeaters
>   Network Interface Cards
>   Network Management
>   Communications Software
>   Networking Software

Abacus Controls, Inc.
80 Readington Rd.
Somerville, NJ 08876
Tel: 908-526-6010
FAX: 908-526-6866
Product Types:
>   UPS

Abbott Systems, Inc.
62 Mountain Rd.
Pleasantville, NY 10570
Tel: 800-552-9157, 914-747-4171
FAX: 914-747-9115
Product Types:
>   Graphics Software
>   GUI Software
>   Personal Information Managers

Accent Data Solutions, Inc.
3270 Seldon Court, Bldg. 1
Fremont, CA 94538
Tel: 510-490-6299
FAX: 510-490-2546

Product Types:
> Scanners
> Accounting Applications
> Terminals

AccuSoft Corporation
Two Westborough Business Park, PO Box 1261
Westborough, MA 01581
Tel: 800-525-3577, 508-898-2770
FAX: 508-898-9662
Product Types:
> Graphics Software
> GUI Software

Adaptec, Inc.
691 S. Milpitas Blvd.
Milpitas, CA 95035
Tel: 800-934-2766, 408-945-8600
FAX: 408-262-2533
Product Types:
> Network Interface Cards
> Sound Boards
> Networking Software
> Peripheral Device Drivers

Adobe Systems, Inc.
1585 Charleston Rd., PO Box 7900
Mountain View, CA 94039-7900
Tel: 800-833-6687, 415-961-4400
FAX: 415-961-3769
Product Types:
> Desktop Publishing Applications
> Peripheral Device Drivers
> Printer Utilities

Advanced Computer Communications (ACC)
10261 Bubb Rd.
Cupertino, CA 95014
Tel: 800-444-7854, 408-864-0600
FAX: 408-446-5234
Product Types:
> Bridges/Routers/Gateways
> Networking Software

Alliance Computer Services
301 E. Wallace, Ste. 110
San Saba, TX 76877
Tel: 800-256-1786, 915-372-5715
FAX: 915-372-5716
Product Types:
    Accounting Applications

Amdahl Corporation
1250 E. Arques Ave., PO Box 3470
Sunnyvale, CA 94088-3470
Tel: 800-538-8460, 408-746-6000
FAX: 408-773-0833
Product Types:
    Mainframes and Supercomputers
    Compilers and Languages
    Networking Software

America Online, Inc.
8619 Westwood Center Dr.
Vienna, VA 22182
Tel: 800-827-6364, 703-448-8700
FAX: 800-827-4595
Product Types:
    Communications Services
    PC Communications Utilities

AMP, Inc.
PO Box 3608
Harrisburg, PA 17105
Tel: 800-488-8459, 717-564-0100
FAX: 717-986-7575
Product Types:
    Modems
    Multiplexors
    Test Equipment

Ancor Communications, Inc.
6130 Blue Circle Dr.
Minnetonka, MN 55343
Tel: 612-932-4000
FAX: 612-932-4037

Product Types:
    Bridges/Routers/Gateways
    Network Interface Cards
    Hubs

Apple Computer, Inc.
20525 Mariani Ave.
Cupertino, CA 95014
Tel: 800-776-2333, 408-996-1010
FAX: 408-996-0275
Product Types:
    Hardware Platforms
    Modems
    Multiplexers
    Network Interface Cards
    Network Management
    Printers
    Electronic Mail

Artisoft, Inc.
2202 N. Forbes Blvd.
Tucson, AZ 85745
Tel: 800-233-5564, 602-670-7100
FAX: 602-670-7101
Product Types:
    Hubs
    Network Interface Cards
    Server/Sharing Units
    Printer Spoolers

Ascend Communications, Inc.
1275 Harbor Bay Pkwy.
Alameda, CA 94502
Tel: 800-ASCEND-4, 510-769-6001
FAX: 510-814-2300
Product Types:
    Bridges/Routers/Gateways

Autodesk, Inc.
111 McGuinness Pkwy.
San Rafael, CA 94903
Tel: 800-879-4233, 415-507-5000

FAX: 415-507-5100
Product Types:
      CAD Software

Banyan Systems, Inc.
120 Flanders Rd.
Westborough, MA 01581-1033
Tel: 800-222-6926, 508-898-1000
FAX: 508-898-1755
Product Types:
      Network Management
      Electronic Mail
      Network Server Software

Black Box Corporation
PO Box 12800
Pittsburgh, PA 15241-0800
Tel: 412-746-5500
FAX: 412-746-0746
Product Types:
      Modems
      Multiplexers
      Test Equipment
      Bridges/Routers/Gateways
      Hubs
      Network Interface Cards
      Network Management
      Fax Boards
      ISDN Adapters

Borland International, Inc.
100 Borland Way
Scotts Valley, CA 95066-3249
Tel: 800-233-2444, 408-431-1000
FAX: 408-431-4122
Product Types:
      Communications Software
      Compilers and Languages
      Design and Testing Software

Cabletron Systems, Inc.
35 Industrial Way
Rochester, NH 03866-5005
Tel: 800-332-9401, 603-332-9400
FAX: 603-337-2211
Product Types:
Bridges/Routers/Gateways
Hubs
Repeaters
Network Interface Cards

Chipcom Corporation
118 Turnpike Rd., Southborough Office Park
Southborough, MA 01772-1886
Tel: 800-228-9930, 508-460-8900
FAX: 508-490-5696
Product Types:
Modems
Multiplexers
Bridges/Routers/Gateways
Hubs
Terminal Servers
Networking Software

Cisco Systems, Inc.
170 W. Tasman Dr.
San Jose, CA 95134-1706
Tel: 800-553-6387, 408-526-4000
FAX: 408-526-4100
Product Types:
Bridges/Routers/Gateways
Hubs
Network Interface Cards
Networking Software

Compaq Computer Corporation
20555 State Hwy. 249
Houston, TX 77070-2698
Tel: 800-345-1518, 713-374-0484
FAX: 713-374-4583
Product Types:
Hardware Platforms
Networking Software

CompuServe, Inc.
5000 Arlington Centre Blvd., PO Box 20212
Columbus, OH 43220
Tel: 800-848-8199, 614-457-8600
FAX: 614-457-0348
Product Types:
> Communications Services

Digital Equipment Corporation
146 Main St.
Maynard, MA 01754-2571
Tel: 800-344-4825, 508-493-5111
FAX: 508-493-8780
Product Types:
> Hardware Platforms
> Modems
> Test Equipment
> Bridges/Routers/Gateways
> Hubs
> Network Interface Cards (NICs)
> Network Management
> Terminal Servers
> Printers
> Electronic Mail
> Networking Software

Eicon Technology Corporation
2196 32nd Ave.
Montreal, QC, CD H8T 3H7
Tel: 800-803-4266, 514-631-2592
FAX: 514-631-3092
Product Types:
> Network Management
> GUI Software
> Networking Software
> ISDN Terminal Adapters

Epson America, Inc.
20770 Madrona Ave., PO Box 2842
Torrance, CA 90503
Tel: 800-289-3776, 310-782-0770
FAX: 310-782-4455

Product Types:
> Hardware Platforms
> Network Interface Cards
> Fax Boards
> Printers

Farallon Computing, Inc.
2470 Mariner Square Loop
Alameda, CA 94501-1010
Tel: 800-425-4141, 510-814-5100
FAX: 510-814-5023
Product Types:
> Bridges/Routers/Gateways
> Hubs
> Network Interface Cards
> Networking Software

Fujitsu Networks Industry, Inc.
1266 E. Main St., Soundview Plaza
Stamford, CT 06902-3546
Tel: 800-446-4736, 203-326-2700
FAX: 203-964-1007
Product Types:
> Help Desk Applications
> Video/Teleconferencing Software

Gandalf Systems Corporation
501 Delran Pkwy.
Delran, NJ 08075-1249
Tel: 800-426-3253, 609-461-8100
FAX: 609-461-4074
Product Types:
> Modems
> Bridges/Routers/Gateways
> Hubs
> Networking Software
> ISDN Adapters

Grand Junction Networks, Inc.
47281 Bayside Pkwy.
Fremont, CA 94538
Tel: 800-747-FAST, 510-252-0726

FAX: 510-252-0915
Product Types:
      Hubs

Hayes Microcomputer Products, Inc.

5835 Peachtree Corners E.
Norcross, GA 30092-3405
Tel: 800-96-HAYES, 800-665-1259 (CD), 404-840-9200
FAX: 404-441-1213
Product Types:
      Modems
      Bridges/Routers/Gateways
      Network Management
      Fax Boards
      Electronic Mail
      Networking Software
      ISDN Adapters

Hewlett-Packard Company

3000 Hanover St.
Palo Alto, CA 94304-1181
Tel: 800-752-0900, 415-857-1501
FAX: 800-333-1917
Product Types:
      Hardware Platforms
      Test Equipment
      Bridges/Routers/Gateways
      Hubs
      Network Interface Cards
      Network Management
      UPS
      Printers
      Electronic Mail
      Networking Software
      ISDN Adapters

Hitachi America, Ltd.

3617 Parkway Lane
Norcross, GA 30092
Tel: 800-446-8820, 404-446-8821
FAX: 404-242-1414
Product Types:
      Network Management

Hughes LAN Systems, Inc.
1225 Charleston Rd.
Mountain View, CA 94043
Tel: 800-395-LANS, 415-966-7300
FAX: 415-966-6161
Product Types:
> Bridges/Routers/Gateways
> Hubs
> Terminal Servers
> Networking Software

IBM
Old Orchard Rd.
Armonk, NY 10504
Tel: 800-426-3333, 914-765-1900
Product Types:
> Hardware Platforms
> Multiplexers
> Bridges/Routers/Gateways
> Hubs
> Network Interface Cards
> Network Management
> Fax Boards
> Software Applications

InSoft, Inc.
4718 Old Gettysburg Rd., Ste. 307, Executive Park W. I
Mechanicsburg, PA 17055
Tel: 717-730-9501
FAX: 717-730-9504
Product Types:
> Video/Teleconferencing Software

Intel Corporation
5200 N.E. Elam Young Pkwy.
Hillsboro, OR 97124-6497
Tel: 800-538-3373, 503-629-7354
FAX: 503-629-7580
Product Types:
> Test Equipment
> Network Interface Cards
> Fax Boards

> Modems
> Software Applications
> Fax software
> Networking Software
> ISDN Adapters

Lantronix

15353 Barranca Pkwy.
Irvine, CA 92718-2216
Tel: 800-422-7055, 714-453-3990
FAX: 714-453-3995
Product Types:
> Bridges/Routers/Gateways
> Terminal Servers

Lotus Development Corporation

55 Cambridge Pkwy.
Cambridge, MA 02142-1295
Tel: 800-343-5414, 617-577-8500
FAX: 617-693-3512
Product Types:
> Office Automation Software
> Spreadsheets
> Electronic Mail Software
> Networking Software

Mass Optical Storage Technologies, Inc.

11205 Knott Ave., Ste. B
Cypress, CA 90630
Tel: 714-898-9400
FAX: 714-373-9960
Product Types:
> Optical Disk Drives

McDATA Corporation

310 Interlocken Pkwy.
Broomfield, CO 80021-3464
Tel: 800-545-5773, 303-460-9200
FAX: 303-465-4996
Product Types:
> Modems
> Multiplexers
> Network Management

MCI Communications Corporation
3 Ravinia Dr.
Atlanta, GA 30346
Tel: 800-825-9675, 404-668-6000
Product Types:
        Communications Services

MicroTest, Inc.
4747 N. 22nd St.
Phoenix, AZ 85016-4700
Tel: 800-526-9675, 602-952-6400
FAX: 602-952-6401
Product Types:
        Test Equipment
        Networking Software

Microsoft Corporation
One Microsoft Way
Redmond, WA 98052-6399
Tel: 800-426-9400, 206-882-8080
FAX: 206-93-MSFAX
Product Types:
        Software Applications
        Electronic Mail Software
        Networking Software

Newbridge Networks, Inc.
593 Herndon Pkwy.
Herndon, VA 22070-5241
Tel: 800-343-3600, 703-834-3600
FAX: 703-471-7080
Product Types:
        Multiplexers
        Bridges/Routers/Gateways
        Network Management
        Networking Software
        ISDN Adapters

Novell, Inc.
122 E. 1700 S.
Provo, UT 84606-6194
Tel: 800-453-1267, 801-429-7000
FAX: 801-429-5155

Product Types:
> Network Operating Systems
> Software Applications
> Electronic Mail

Racal-Datacom, Inc.
1601 N. Harrison Pkwy.
Sunrise, FL 33323-2802
Tel: 800-RACAL-55, 305-846-4811
FAX: 305-846-4942
Product Types:
> Modems
> Multiplexers
> Bridges/Routers/Gateways
> Hubs
> Network Management
> Terminal Servers
> Networking Software

Retix
2401 Colorado Ave., Ste. 200
Santa Monica, CA 90404-3563
Tel: 800-255-2333, 310-828-3400
FAX: 310-828-2255
Product Types:
> Multiplexors
> Bridges/Routers/Gateways
> Hubs
> Software Applications
> Electronic Mail
> Networking Software

StrataCom, Inc.
1400 Parkmoor Ave.
San Jose, CA 95126
Tel: 800-767-4479, 408-294-7600
FAX: 408-999-0115
Product Types:
> Multiplexers
> Network Management
> Routers

Sun Microsystems Computer Corporation
2550 Garcia Ave.
Mountain View, CA 94043-1100
Tel: 800-821-4643, 415-960-1300
FAX: 415-969-9131
Product Types:
> Hardware Platforms
> Network Interface Cards
> Software Applications
> Networking Software

SynOptics Communications, Inc.
4401 Great America Pkwy., PO Box 58185
Santa Clara, CA 95054-8185
Tel: 800-PRO-NTWK, 408-988-2400
FAX: 408-988-5525
Product Types:
> Bridges/Routers/Gateways

TASC, Inc.
2555 University Drive
Fairborn, OH 45324
Tel: 513-426-1040
FAX: 617-426-8888
Product Types:
> Software Applications
> Document Management

Toshiba America Information Systems, Inc.
9740 Irvine Blvd., PO Box 19724
Irvine, CA 92713-9724
Tel: 800-334-3445, 714-583-3000
FAX: 714-583-3645
Product Types:
> Hardware Platforms
> Network Interface Cards
> Fax Boards
> Modems

UB Networks, Inc.
PO Box 58030, 3900 Freedom Circle
Santa Clara, CA 95052-8030
Tel: 800-777-4LAN, 408-496-0111

FAX: 408-970-7300
Product Types:
> Bridges/Routers/Gateways
> Hubs
> Network Interface Cards
> Network Management
> Terminal Servers
> Networking Software

Unisys Corporation

PO Box 500
Blue Bell, PA 19424-0001
Tel: 800-874-8647, 215-986-4011
FAX: 215-986-3170
Product Types:
> Hardware Platforms
> Software Applications
> Communications Software

U.S. Robotics, Inc.

8100 N. McCormick Blvd.
Skokie, IL 60076-2999
Tel: 800-USR-CORP, 708-982-5010
FAX: 708-933-5800
Product Types:
> Modems
> Network Management
> Fax Boards
> Networking Software

Wang Laboratories, Inc.

600 Technology Park Dr.
Billerica, MA 01821-4130
Tel: 800-225-0654, 508-967-5000
FAX: 508-967-0828
Product Types:
> Hardware Platforms

Wireless Networking Products and Services

As part of a wireless network implementation, you will certainly need to decide which products to use. This appendix contains a high-level description of wireless networking products and services to aid you in making a decision. A separate section is devoted to each of the following:

○ Wireless LAN products

○ Wireless MAN products

○ Wireless WAN products and services

The sections contain tables identifying specific products and comparing their attributes. Based on your network's requirements, scan the product lists and identify potential products; then obtain more detailed product information from the applicable vendors. Each of the tables identifies the products' vendor. Contact information for these vendors can be found in Appendix A, "Wireless Networking Vendors and Service Companies."

Wireless LAN Products

Table B.1 contains descriptions of radio and infrared-based wireless LAN network interface cards and access points. These products provide connectivity among clients and servers and access to wired ethernet networks within a local area, such as a room, warehouse, or building. All given ranges are approximate for semi-open areas, such as offices with partitions and walls. Longer ranges are possible in open areas, such as warehouses and auditoriums. In addition, most of these products offer access points that extend the range by forming a multiple cell configuration with wired ethernet or token-ring networks as a backbone.

Table B.1

Wireless LAN Products.

Product Name	Company	Interface Protocol	Wireless Technology	Data Rate	Range	Comments
AIRLAN	Solectek	ISA, PCMCIA II, Parallel port	Spread spectrum radio (ISM Band)	2 Mbps	200 feet	AIRLAN uses WaveLAN technology.
Altair	Motorola	IEEE 802.3 Ethernet	18 Ghz narrowband radio	5.7 Mbps	100 feet	Requires FCC licensing; uses a centralized wireless hub.
ARLAN	Aironet	ISA, PCMCIA	Spread spectrum radio (ISM Band)	2 Mbps	200 feet	Radios use direct sequence spread spectrum; Aironet's ARLAN 630 and ARLAN 631 access points

Product Name	Company	Interface Protocol	Wireless Technology	Data Rate	Range	Comments
						provide multiple cell configuration via ethernet and token ring respectively.
Aviator Wireless Network	Nova Counter-Electronics	Parallel port	Spread spectrum radio (ISM Band)	150 Kbps	60 feet	
BreezeNET	BreezeCOM	ISA, PCMCIA II	Spread spectrum radio (ISM Band)	3 Mbps	200 feet	Radios use frequency hopping spread spectrum; Access Point AP-10 provides multiple cell configuration.
Collaborative	Photonics	ISA, PCMCIA II, parallel port	Diffused infrared	1 Mbps	25 feet	Photonic's Collaborative

continues

Wireless LAN Products Continued

Product Name	Company	Interface Protocol	Wireless Technology	Data Rate	Range	Comments
						Access Point provides multiple cell configuration via ethernet.
Cooperative	Photonics	Apple LocalTalk	Diffused infrared	230.4 Kbps	25 feet	Photonic's Cooperative Access Point provides multiple cell configuration via ethernet.
CruiseLAN	Zenith Data Systems	ISA, PCMCIA II	Spread spectrum radio (ISM Band)	1.6 Mbps	200 feet	Zenith's CruiseLAN/ Access Point provides multiple cell configuration via ethernet; features software encryption.

Product Name	Company	Interface Protocol	Wireless Technology	Data Rate	Range	Comments
FreePort	Windata, Inc.	IEEE 802.3 Ethernet	Spread spectrum radio (ISM Band)	5.7 Mbps	200 feet	Uses a centralized wireless hub.
GoPrint	AeroComm	Centronics parallel port	Spread spectrum radio (ISM Band)	1 Mbps	200 feet	
Grouper	Digital Ocean	Apple	Spread spectrum radio (ISM Band)	2 Mbps		Supports Apple product line; Apple Newton PDA; uses WaveLAN technology.
InfraLAN	InfraLAN Technologies	IEEE 802.5 Token Ring	Point-to-point infrared	4 or 16 Mbps	80 feet	
Infrared Wireless Adapter	IBM	ISA, PCMCIA II	Diffused infrared	1 Mbps	30 feet	
Netwave	Xircom	PCMCIA II	Spread spectrum radio (ISM Band)	1 Mbps	200 feet	Radios use frequency hopping spread spectrum;

continues

Wireless LAN Products Continued

Product Name	Company	Interface Protocol	Wireless Technology	Data Rate	Range	Comments
						Antenna is integrated within the PCMCIA adapter; Netwave Access Point for Ethernet provides multiple cell configuration; incorporates authentication and data scrambling features.
Novell Embedded Systems Technology (NEST)	Novell, Inc.	(Unknown)	Carrier currents over electrical power lines	2 Mbps	Length of power line circuit	This product is projected for release in 1997.

Product Name	Company	Interface Protocol	Wireless Technology	Data Rate	Range	Comments
RangeLAN	Proxim	ISA, PCMCIA II	Spread spectrum radio (ISM Band)	1.6 Mbps	200 feet	Proxim's access point provides multiple cell configuration via ethernet.
SpextrixLite	Spectrix	PCMCIA II, RS-232, TTL, RS-485/422	Diffused infrared	4 Mbps	50 feet	SpectrixLite's Wireless LAN Base Station provides multiple cell configuration.
WaveLAN	Lucent	ISA, PCMCIA II	Spread spectrum radio (ISM Band)	2 Mbps	200 feet	Radios use direct sequence spread spectrum; Lucent's Netwave Access Point for Ethernet provides multiple cell configuration

continues

Wireless LAN Products Continued

Product Name	Company	Interface Protocol	Wireless Technology	Data Rate	Range	Comments
						via ethernet; a data encryption chip is an available option for added security.
Wireless LAN	IBM	ISA, PCMCIA II	Spread spectrum radio (ISM Band)	1 Mbps	200 feet	Radios use frequency hopping spread spectrum; IBM's access point provides multiple cell configuration via ethernet and token ring; continuous data encryption.

Wireless MAN Products

Table B.2 contains descriptions of radio and infrared-based wireless MAN products. These products offer outdoor point-to-point connectivity between buildings and other structures. The ranges given in the table are approximate—the true ranges depend on weather and the presence of physical obstructions and interference.

Table B.2

Wireless MAN Products.

Product Name	Company	Interface Protocol	Wireless Technology	Data Rate	Range	Comments
AIRLAN/ Bridge Plus	Solectek	IEEE 802.3 Ethernet	Spread spectrum radio (ISM Band)	2 Mbps	3 miles	
AirLink	Cylink	IEEE 802.3 Ethernet	Spread spectrum radio (ISM Band)	1872 64-byte pps	15 miles	Radios use direct sequence spread spectrum.
AirPort	Windata, Inc.	IEEE 802.3 Ethernet	Spread spectrum radio (ISM Band)	5.7 Mbps	1 mile	Radios use direct sequence spread spectrum.

continues

Wireless MAN Products Continued

Product Name	Company	Interface Protocol	Wireless Technology	Data Rate	Range	Comments
Altair VistaPoint	Motorola	IEEE 802.3 Ethernet	18 GHz narrowband radio	5.3 Mbps	4,000 feet	
ARLAN	Aironet	IEEE 802.3 Ethernet	Spread spectrum radio (ISM Band)	2 Mbps	3 miles	Radios use direct sequence spread spectrum.
BreezeLINK	BreezeCOM	T1, E1, V.35, RS-530, X.21	Spread spectrum radio (ISM Band)	56–2048 Kbps	20 miles	Radios use frequency hopping spread spectrum.
EthAirBridge	LANAIR	IEEE 802.3 Ethernet	Spread spectrum radio (ISM Band)	300 Kbps	15 miles	Radios use frequency hopping spread spectrum.
FASTWAVE 850	Southwest Microwave	IEEE 802.3 Ethernet	23 GHz narrowband radio	10 Mbps	3 miles	Supports full-duplex (20 Mbps) ethernet.

Product Name	Company	Interface Protocol	Wireless Technology	Data Rate	Range	Comments
FASTWAVE 950	Southwest Microwave	Fractional T1, T1, 4T1, or IEEE 802.3 Ethernet	23 GHz narrowband radio	10 Mbps	10 miles	Supports full duplex (20 Mbps) ethernet.
Intersect Remote Bridge	Persoft	IEEE 802.3 Ethernet or IEEE 802.5 Token Ring	Spread spectrum radio (ISM Band)	2 Mbps	3 miles	
OmniBeam 2000	Laser Communications	IEEE 802.3 Ethernet, IEEE 802.5 Token Ring, T1, E1, V.35	laser (820 nm)	Line Speed	3,940 feet	
OmniBeam 4000	Laser Communications	E3, T3, SONET-1, SONET-3, ATM-52, 100BaseT, FDDI	laser (820 nm)	34–155 Mbps	Depends on data rate	Features an optical fiber interface.

continues

Wireless MAN Products Continued

Product Name	Company	Interface Protocol	Wireless Technology	Data Rate	Range	Comments
RAN 128/900	Multipoint Networks	V.35	820–960 MHz narrowband radio	128 Kbps	30 miles	Other RAN products are available that operate at lower data rates.
RangeLINK	Proxim	IEEE 802.3 Ethernet	Spread spectrum radio (ISM Band)	1.6 Mbps	3 miles	
Wireless Link	SpreadNet	IEEE 802.3 Ethernet	Spread spectrum radio (ISM Band)	2 Mbps	25 miles	

Wireless WAN Products and Services

Table B.3 contains descriptions of various wireless WAN products. These products offer wide area (with the United States and worldwide) connectivity between portable computers and the Internet, corporate networks, and wireless WAN service provider networks.

Table B.3

Wireless WAN Products.

Product Name	Company	Interface Protocol	Wireless Technology	Data Rate	Range	Comments
MCC-550C Remote Station	Meteor Communications Corporation	RS-232, Parallel	Meteor burst communications (40–50 MHz)	9,600 bps	1000 miles	Interfaces with the MCC-520B Meteor Burst Master Station.
MobileSat LYNXX Transportable Inmarsat-B Earth Station	California Microwave	RS-232, RS-422/449, V.35	Satellite	64 Kbps	Worldwide	

continues

Wireless WAN Products Continued

Product Name	Company	Interface Protocol	Wireless Technology	Data Rate	Range	Comments
Personal Messenger 100C Wireless Modem Card	Motorola	PCMCIA, RS-232	CDPD	19.2 Kbps	Depends on the coverage area of the selected CDPD service provider	Interfaces with CDPD network service.
Personal Messenger 100D Wireless Modem Card	Motorola	PCMCIA	Packet radio (ARDIS)	20 Kbps	Depends on ARDIS coverage area	Interfaces with the ARDIS packet radio network service.
Ricochet	Metricom	RS-232	Packet radio (Ricochet)	28.8 Kbps	Depends on coverage of Metricom's Ricochet service	

Product Name	Company	Interface Protocol	Wireless Technology	Data Rate	Range	Comments
Wireless Modem for ARDIS	IBM	PCMCIA	Packet radio (ARDIS)	19.2 Kbps	Depends on the coverage area of the selected CDPD service provider	Interfaces with the ARDIS packet radio network service.
Wireless Modem for Cellular/CDPD	IBM	PCMCIA	CDPD	19.2 Kbps	Depends on the coverage area of the selected CDPD service provider	Interfaces with CDPD network service.
Wireless Modem for Mobitex (RAM Mobile Data)	IBM	PCMCIA	Packet radio (Mobitex)	8 Kbps	Depends on the RAM Mobile Data coverage area	Interfaces with the RAM Mobile Data packet radio network service.

Wireless WAN Services

Table B.4 contains descriptions of wireless WAN services, which require the use of a wireless WAN interface as described in Table B.3. Wireless WAN services offer wide area wireless public data networking for mobile users.

Table B.4

Wireless WAN Services.

Service Name	Company	Wireless Technology	Data Rate	Range	Comments
ARDIS	ARDIS	Packet radio	20 Kbps	Most of the United States	Requires a wireless modem.
AT&T Wireless Packet Data Service	AT&T Wireless Services	CDPD	19.2 Kbps	Worldwide	Requires a CDPD modem.
RAM Mobile Data	RAM Mobile Data	Packet radio	20 Kbps	Most of the United States	Requires a wireless modem.
Ricochet	Metricom	Packet radio	28.8 Kbps	Selected areas of the United States	Requires a wireless modem.

Standards Organizations and Industry Groups

This appendix contains a description of the leading standards organizations and industry groups.

ACM

ACM was founded in 1947 as an international scientific and educational organization dedicated to advancing the art, science, engineering, and application of information technology. ACM serves both professional and public interests by fostering the open interchange of information and by promoting the highest professional and ethical standards. ACM membership today consists of some 80,000 men and women who are largely practitioners, developers, researchers, educators, engineers, and managers, all with significant interest in the creation and application of information technologies.

> ACM
> One Astor Plaza
> 1515 Broadway
> New York, NY 10036
> 212-869-7440
> 212-944-1318 (fax)
> ACMHELP@acm.org

American National Standards Institute (ANSI)

The American National Standards Institute (ANSI) is a privately funded federation of leaders representing both the private and public sectors, responsible for coordinating the U.S. voluntary consensus standards system. ANSI, organized in 1918, is made up of manufacturing and service businesses, professional societies and trade associations, standards developers, academia, government agencies, and consumer and labor interests, all working together to develop voluntary national consensus standards. ANSI provides U.S. participation in the international standards community as the sole U.S. representative to the two major non-treaty international standards organizations: the International Organization for Standardization (ISO) and, through the U.S. National Committee, the International Electrotechnical Commission (IEC).

American National Standards Institute (ANSI)
11 West 42nd Street
New York, NY 10036
212-642-4948
212-398-0023 (fax)

Asociación de Técnicos de Informática (ATI)

The goals of ATI are to improve, promote, and defend the work of computer professionals, especially within Spain, Europe, and Latin America.

Asociación de Técnicos de Informática
Via Laietana 41, 1° 1°
08003-BARCELONA
34 3 412 52 35
34 3 412 77 13 (fax)
secregen@ati.es

Association for Systems Management (ASM)

The Association for Systems Management (ASM) is a non-profit association of information systems professionals. ASM's focus is on the management of information systems and information technology. Their mission is to provide an environment for professional development, continuing education, and networking for information systems professionals.

> 800-203-3657
> 74431.3442@compuserve.com

Association for Women in Computing (AWC)

The Association for Women in Computing (AWC) is a non-profit organization dedicated to promoting the advancement of women in computing professions. The AWC was founded as a non-profit organization by 15 women in Washington, D.C. in 1978. AWC is dedicated to the advancement of women in the computing fields, business, industry, science, education and government. AWC's purpose is to provide opportunities for professional growth through networking and programs covering technical and career-oriented topics. AWC promotes awareness of issues affecting women in the computing industry and furthers the professional development and advancement of women in computing.

> Association for Women in Computing
> 41 Sutter Street, Suite 1006
> San Francisco, CA 94104
> 415-905-4663
> awc@acm.org

ATM Forum

The ATM Forum is an international non-profit organization formed with the objective of accelerating the use of ATM (Asynchronous

Transfer Mode) products and services through a rapid convergence of interoperability specifications. In addition, the Forum promotes industry cooperation and awareness.

> The ATM Forum
> 2570 W. El Camino Real, Suite 304
> Mountain View, CA 94040-1313
> 415-949-6700
> 415-949-6705 (fax)
> info@atmforum.com

Australian Computer Society (ACS)

The Australian Computer Society (ACS) is the primary professional association in Australia for those in the computing and information technology fields. The ACS was established in 1966 as a result of the merger of then-existing State-based computer societies.

> Australian Computer Society
> PO Box 319
> DARLINGHURST NSW 2010
> http://www.acs.org.au/

Black Data Processing Associates (BDPA)

The mission of Black Data Processing Associates (BDPA) is to sustain a network of information technology professionals that is a positive influence in the information processing industry—a network that shares information, provides education, and performs community service. On a local and national level, the primary objective of BDPA is to accumulate a pool of information processing knowledge and business expertise with the intention of utilizing those resources to support the interest of individuals examining the field of information processing as a career or business opportunity and provide a forum for the development of interests and skills of the minority community as a whole.

Black Data Processing Associates
1250 Connecticut Avenue NW, Suite 610
Washington, DC 20036-2603
800-727-BDPA
202-775-4301
info@bdpabac.com

British Computer Society

The British Computer Society
1 Sanford Street
Swindon, Wiltshire SN1 1HJ
+44 01793 417417
+44 01793 480270 (fax)
bcshq@bcs.org.uk

CDPD Forum

The CDPD Forum is a not-for-profit association of companies dedicated to the development and growth of the cellular data industry and, in particular, to the standardization and effective use of cellular digital packet data (CDPD) technology.

The CDPD Forum was chartered and incorporated in April 1994 as a trade association of wireless data service providers, equipment manufacturers, software developers, and information providers who shape this technology and support the development of the CDPD commercial marketplace. All Forum memberships are corporate, not individual. Membership in the CDPD Forum is open to all companies that develop, deliver, or use CDPD products or services. Currently, there are more than 90 member companies, nine of which sit on the board of directors.

CDPD Forum c/o Smith, Bucklin & Associates, Inc.
401 North Michigan Ave.
Chicago, IL 60611-4267
800-335-CDPD
312-644-6610
312-321-6869 (fax)
info@forum.cdpd.net

Computing Research Association (CRA)

The Computing Research Association (CRA) is an association of North American academic departments of computer science and computer engineering, industrial laboratories engaging in basic computing research, and affiliated professional societies. CRA's mission is to represent and inform the computing research community and support and promote its interests. CRA seeks to strengthen research and education in the computing fields, expand opportunities for women and minorities, and improve public and policy maker understanding of the importance of computing and computing research in our society.

> Computing Research Association
> 1875 Connecticut Avenue NW, Suite 718
> Washington, DC 20009
> 202-234-2111
> 202-667-1066 (fax)
> info@cra.org

Data Processing Management Association (DPMA)

Founded in 1951 by a progressive group of accounting department data processors, Data Processing Management Association (DPMA) was initially called the National Machine Accountants Association. The present name was adopted in 1962. DPMA stands as the leader among organizations representing information systems (IS) professionals. With 9,000 members and 5,000 student members throughout the U.S. and Canada, DPMA's mission is to promote the effective, responsible management of information technology to the benefit of its members, their employers, and society.

> DPMA
> 505 Busse Hwy
> Park Ridge, IL 60068-3191
> 847-825-8124
> 847-825-1693 (fax)
> 70430.35@compuserve.com

Frame Relay Forum

The Frame Relay Forum is an association of corporate members comprised of vendors, carriers, users, and consultants committed to the implementation of Frame Relay in accordance with national and international standards.

> The Frame Relay Forum
> 303 Vintage Park Drive
> Foster City, CA 94404-1138
> 415-578-6980
> 415-525-0182

Infrared Data Association

Infrared Data Association (IrDA) was established in 1993 to set and support hardware and software standards that create infrared communications links. The Association's charter is to create an interoperable, low-cost, low-power, half-duplex, serial data interconnection standard that supports a walk-up, point-to-point user model and is adaptable to a wide range of applications and devices. IrDA standards support a broad range of computing, communications, and consumer devices.

International in scope, IrDA is a non-profit corporation headquartered in Walnut Creek, California and led by a Board of Directors that represents a voting membership of more than 150 corporate members worldwide. As a leading high technology standards association, IrDA is committed to developing and promoting infrared standards for the hardware, software, systems, components, peripherals, communications, and consumer markets.

> Infrared Data Association
> P.O. Box 3883
> Walnut Creek, CA 94598
> 510-943-6546
> 510-934-5241 (fax)
> daphne@irda.org

Institute of Electrical and Electronic Engineers (IEEE)

The Institute of Electrical and Electronic Engineers (IEEE) is a non-profit professional organization founded by a handful of engineers in 1884 for the purpose of consolidating ideas dealing with electro-technology. In the last 100-plus years, IEEE has maintained a steady growth. Today, the IEEE, which is based in the USA, has over 320,000 members located in 150 countries. The IEEE consists of 35 individual societies, including the Communications Society, Computer Society, and Antennas and Propagation Society, to name just a few. The Computer Society is the official sponsor of the IEEE P802.11 Working Group.

The IEEE plays a significant role in publishing technical works, sponsoring conferences and seminars, accreditation, and standards development. IEEE has published nearly 700 active standards publications, half relating to power engineering and most others dealing with computers. The IEEE standards development process consists of 30,000 volunteers (who are mostly IEEE members) and a Standards Board of 32 people. In terms of LANs, IEEE has produced some very popular and widely used standards. For example, the majority of LANs in the world use network interface cards based on the IEEE 802.3 (ethernet), IEEE 802.5 (token ring), and the IEEE 802.11 (wireless LAN) standards.

IEEE
445 Hoes Lane
Piscataway, NJ 08855-0459
800-678-IEEE
212-705-7900
http://www.ieee.org/index.html
member.services@ieee.org

International Organization for Standardization (ISO)

The International Organization for Standardization (ISO) is a worldwide federation of national standards bodies. ISO is a non-governmental organization established in 1947. The mission of ISO is to promote the development of standardization and related activities with a view toward facilitating the international exchange of goods and services and developing cooperation in the spheres of intellectual, scientific, technological, and economic activity. ISO's work results in international agreements that are published as International Standards.

> International Organization for Standardization
> 1, rue de Varembé
> Case postale 56
> CH-1211 Genève 20
> Switzerland
> + 41 22 749 01 11
> central@isocs.iso.ch

International Telecommunication Union (ITU)

The International Telecommunication Union (ITU) is an intergovernmental organization within which the public and private sectors cooperate for the development of telecommunications. The ITU was founded in Paris in 1865 as the International Telegraph Union. ITU took its present name in 1934 and became a specialized agency of the United Nations in 1947.

The ITU adopts international regulations and treaties governing all terrestrial and space uses of the frequency spectrum, as well as the use of the geostationary-satellite orbit, within which countries adopt their national legislation. The ITU also develops standards to facilitate the interconnection of telecommunication systems on a worldwide scale, regardless of the type of technology used. Spearheading telecommunications development on a world scale, the ITU fosters the development of tele-communications in developing countries by establishing medium-term development policies and strategies in consultation with other partners in the sector; and by providing specialized technical assistance in the areas of telecommunication policies, the choice and transfer of technologies, management, financing of investment projects and mobilization of resources, the installation and maintenance of networks, and the management of human resources, research, and development.

International Telecommunication Union (ITU)
Place des Nations
1211 Geneva 20
Switzerland
+41 22 730-6666
+41 22 730-5337 (fax)
helpdesk@itu.ch

Internet Engineering Task Force (IETF)

The Internet Engineering Task Force (IETF) provides a forum for working groups to coordinate technical developments of new protocols. Its most important function is the development and selection of standards within the Internet protocol suite. The IETF began in January 1986 as a forum for technical coordination by contractors for the U.S. Defense Advanced Projects Agency (DARPA), working on the ARPANET, U.S. Defense Data Network (DDN), and the Internet core gateway system. Since that time, the IETF has grown into a large, open international community of network designers, operators, vendors, and researchers concerned with the evolution of the Internet architecture and the smooth operation of the Internet.

http://www.ietf.cnri.reston.va.us/home.html

Internet Society

The Internet Society is a non-governmental International organization for global cooperation and coordination for the Internet and its internetworking technologies and applications. The Society's members reflect the breadth of the entire Internet community and consist of individuals, corporations, non-profit organizations, and government agencies. Its principal purpose is to maintain and extend the development and availability of the Internet and its associated technologies and applications and is a means of enabling organizations, professions, and individuals worldwide to more effectively collaborate, cooperate, and innovate in their respective fields and interests.

The Internet Society was announced in June 1991 at an international networking conference in Copenhagen and was brought into existence in January 1992 by a worldwide cross-section of individuals and organizations. These people recognized that the Society was a critical component necessary to evolve and globalize the Internet and internet technologies and applications, and to enhance their availability and use on the widest possible scale.

Internet Society International Secretariat
12020 Sunrise Valley Drive
Suite 210
Reston, VA 22091
USA
703-648-9888
703-648-9887 (fax)
isoc@isoc.org

Mobile Management Task Force (MMTF)

The Mobile Management Task Force (MMTF) is an industry group that promotes new management standards that specifically address the concerns of network administrators who must manage mobile computer users. The MMTF was first formed by Epilogue Technology Corporation and Xircom, Inc. Current MMTF member companies include IBM Networking Division, Lannair, Motorola, National Semiconductor Corp., and Zenith Data Systems. The joint aim of the group is to define and address the needs of mobile computer users, which are inherently different from those of desktop computer users. As defined by the MMTF charter, the group's aim is to "identify the administrative needs of laptop workstation users, mobile computer users, palmtop users, and others who need reliable access to computer networks on a sporadic basis. The concerns of the Mobile Management Task Force will include (but are not limited to) managing on-demand access to local area networks, dial-up network access, wireless LAN communications, and related administration issues unique to the needs of mobile computer users."

MMTF c/o Epilogue Technology Corporation
10501 Montgomery N.E., Ste. 250
Albuquerque, NM 87111
mmtf-request@epilogue.com

Object Management Group

The Object Management Group (OMG) is a non-profit consortium dedicated to promoting the theory and practice of object technology (OT) for the development of distributed computing systems. OMG was formed to help reduce the complexity, lower the costs, and hasten the introduction of new software applications. OMG's goal is to provide a common architectural framework for object-oriented applications based on widely available interface specifications. OMG was founded in May 1989 by eight companies: 3Com Corporation, American Airlines, Canon, Inc., Data General, Hewlett-Packard, Philips Telecommunications N.V., Sun Microsystems, and Unisys Corporation. In October 1989 OMG began independent operations as a non-profit corporation.

> Object Management Group, Inc.
> Framingham Corporate Center
> 492 Old Connecticut Path
> Framingham, MA 01701
> 508-820-4300
> 508-820-4303 (fax)
> http://www.omg.org/

Portable Computer and Communications Association (PCCA)

The Portable Computer and Communications Association (PCCA) was established in 1992 to advance the portable computing industry.

> PCCA
> P.O. Box 924
> Brookdale, CA 95007
> 408-338-0924
> pcca@mcimail.com

Wireless Opportunities Coalition (WOC)

The Wireless Opportunities Coalition is a diverse group of organizations and companies dedicated to preserving and expanding the opportunities for growth in the wireless industry. The Coalition's primary focus is to support the development, manufacture, and use of wireless communications and related devices that are not licensed by the Federal Communications Commission (FCC) but are regulated under "Part 15" of the FCC's rules. The Wireless Opportunities Coalition is working to communicate—to the FCC, Congress, and the public—the importance of preserving the use of these low-cost, unlicensed devices on the public airwaves.

http://policy.net/wireless/

Wireless Networking in Healthcare

NOTE

The authors of this section are Saleem Desai, M.D., and Balasubramanian Ramachandran from Medical Communication Systems (MCS), Inc. MCS is a medical informatics company based in Boston, MA, and specializes in the application of emerging technology to enhance and expedite healthcare delivery.

Medicine, like many scientific activities, cannot be practiced effectively without accurate and timely information, such as information about patients and their problems, appropriate care giving procedures, and their benefits and limitations. Today, the vast bulk of the information cannot be found online, but in one hard copy form or another, such as paper based patient records derived from provider (patient encounters, film-based X-rays, and scans from diagnostic procedures). All are critical components of what, in aggregate, is called the patient record. Computerizing the patient record and enabling wireless retrieval is an excellent way to reduce administrative costs and deliver better healthcare facilities.

Wireless technology in the form of cordless, cellular telephony and paging has enhanced the way of life of the common end user. The benefits, however, are not limited to private users. Using wireless technology in the medical setting can greatly improve the productivity of healthcare providers and the accuracy of diagnoses by permitting the retrieval of patient related data by physicians in remote locations.

The wireless application to healthcare grew out of a combination of existing market needs. As highly mobile professionals, physicians require immense amounts of data to make critical, life-saving decisions in a timely fashion. Recent trends in healthcare, such as cost-containment, susceptibility to malpractice suits, and the need to access large amounts of information, have forced physicians to spend more time conducting administrative work, rather than actual patient care. These trends increase the necessity of a wireless network to facilitate the transmission and access of the computerized patient records from one medical facility to another within a particular geographic area. A wireless handheld device carried by medical personnel will enable him or her to retrieve and relay patient-related data in an expeditious manner. This appendix primarily focuses on the design and implementation issues surrounding wireless personal communication systems in the healthcare environment, including related network architecture issues, traffic, time, and radio resource management.

Benefits of a Wireless Network in Healthcare

Hospital facilities require a great deal of mobility due to the medical personnel constantly being in motion. In such a setting, a wireless local area network (LAN) can provide many benefits for healthcare. By providing wireless portable terminals to doctors, nurses, and hospital administration staff, for example, patient data can be accessed from practically anywhere within the hospital building, and a complete record of prescriptions and medicines can be tracked to the patient's bedside. Such a system can increase the efficiency of medical personnel, thereby reducing healthcare costs. Pertinent medical information is available in real time from multiple sources. Interconnecting the medical facilities enables the transfer of medical and patient images from X-ray scans, consumer health information, and decision analyses.

Central Computerized Patient Record (CCPR)

Within a hospital, the information obtained from multiple sources can be compiled, analyzed, and stored in the central repository—Central Computerized Patient Record (CCPR). This repository is made available in real time by wireless means to medical personnel.

The CCPR communications layer also connects wireless clients that communicate directly with each other. This layer enables medical experts in different fields to deliver quality healthcare by exchanging multimedia information in real time.

Patient case information is generally stored in different databases depending on information content. Medical facilities have numerous modules including radiology, pharmacy, billing, and laboratory that represent different segments of the medical process. These modules have separate databases to store the relevant information, but in the past there was no interaction between these databases to provide a complete picture to the physician.

CCPR provides a central Database Management System (DBMS) that integrates these various modules with a client server methodology. The CCPR design enables any additional module to be plugged in easily by implementing a standard API.

CCPR can be further subdivided into the following subsystems :

○ Graphical user interface (GUI) with CCPR clients.

○ CCPR server that extracts information from the repository and interacts with other servers.

○ Database servers for all modules.

The GUI is simple, user-friendly, and portable across PC platforms and handheld devices. In a healthcare environment, the integration is highly advantageous because the CCPR client establishes a client process that communicates with the servers of other modules. The CCPR client, for example, interacts with the Radiology server module to extract chest X-rays on a given case. The GUI would then display this multimedia information to the medical personnel in an appropriate manner determined by the data type.

Supplementary modules, such as a medical knowledge base, can be incorporated into the system. The medical knowledge base module provides an access path for medical facts or information. This information can be used by medical experts for prognosis, diagnosis, and management. In effect, the CCPR provides a timely update of patient records and an access path for medical facts and information.

Architecture of a Hospital-based Wireless PCs

Mobility and service portability problems must be addressed when using an integrated network. Medical personnel, for example, must be able to extract all necessary patient information through a single directory number. Terminal mobility problems are accommodated by using a portable terminal with wireless access to a fixed base station. Personal mobility can be accommodated through wireless access by using a compact identity card equipped with a wired or wireless access.

Terminal Mobility (TM) and Handover

Terminal Mobility (TM) offers subscribers the ability to roam from one radio coverage area to another. The network, then, must keep track of the movements of the mobile terminal at all times. Another key issue in any wireless system is the ability to continue an existing call as the user moves from one radio coverage to another; this is known as system handover. In a healthcare environment, efficient handover must be performed because the physician is extremely mobile and the continuation of the call from one radio coverage to another is critical. Some wireless systems prioritize handovers over originating calls, but in a hospital environment, both tasks are equally important.

If the user is walking, then the handover is relatively slow. If the user is moving in a vehicle, however, handover has to be faster. In deciding when to perform a handoff, it is important to ensure that the drop in the signal level is not due to momentary fading and the terminal is actually moving away from the base station. To achieve this, a certain threshold level of the signal is fixed depending on the traffic behavior. The base station then monitors the signal level for a certain period of time before a handover is initiated. This period is known as the dwell time and varies depending on the call statistics. This running measurement of signal strength (RSSI) should be optimized so that unnecessary hand-offs are avoided while ensuring that the required hand-offs are completed before a call is terminated due to a low signal level.

In effect, performing handover is a time critical issue. The continuity of the call is ensured by handover when the radio resource changes within the cell (intra-cell handover) or between cells (inter-cell handover). The handover process involves three successive phases: namely measurement, initiation, and handover control. The handover control process depends strictly on the particular radio access method chosen and channel allocation schemes.

Personal Mobility

Personal mobility (PM) relates to a user carrying a personal subscription identity (personal telecommunication number) rather than a terminal. When a caller dials the number, it is the network's responsibility to route that call to the terminal of the subscriber's choice. This could be done through either a manual entry or automatic registration by using a personal identity module.

Using this personal subscription identity, the user can access services from any terminal, whether it is in a fixed or mobile communication network. This is an extremely important and useful requirement in a hospital. The medical personnel are constantly moving between the emergency rooms, diagnostic centers, and other places of work. They also have to spend a considerable amount of time attending to administrative and patient related matters. Because of these considerations, ready access of data from any terminal of their choice is highly advantageous. To achieve this kind of personal and terminal mobility, a database structure and definition of the parameters are required to distinguish between personal mobility and terminal mobility. To enable the free combination of user and mobile terminal related data, maintaining two separate databases is essential. Authentication of the mobile terminals and the users have to be done separately in such a case.

Radio Resource Management

The frequency spectrum is a limited natural resource and efficient utilization of the scarce spectrum available in any wireless system is a challenging design task. Therefore, frequency management is an important issue. The same frequency can be reused repeatedly in different geographical locations. Various locations using the same frequency are called cochannel cells.

The minimum separation required between two nearby cochannel cells is based on specifying a tolerable cochannel interference that is determined by the required carrier-to-interference C/I ratio. As the cochannel separation distance increases, the interference correspondingly decreases. At the same time, however, there is also a decrease in the capacity. Thus, there is a trade-off between the re-use factor and the tolerable interference.

Channel Allocation Schemes

Channel assignment strategies can be classified essentially as fixed channel allocation (FCA) and the dynamic channel allocation (DCA). In a fixed assignment scheme, the number of channels allocated to a particular cell is fixed, making management easier. An inherent disadvantage of the FCA scheme, however, is that once all the channels in a cell have been allocated, any other call attempt at that site is blocked. The dynamic channel allocation method relies on a call-by-call optimization where all the channels are managed by the mobile switching center (MSC). For each call attempt, channels are assigned to the user on a service-on-demand basis by the network. Here, the blocking probability of a call is less than the fixed channel scheme, but the management load of the network increases correspondingly. Other hybrid schemes exist where part of the total number of channels available is allocated to each cell and the remaining channels are in control of the Mobile Switching Center (MSC). The MSC allocates them on demand to the user. Also, there is a trade-off involved between the management load on the network and the specific channel allocation scheme. In a healthcare environment, maintaining the continuity of data is extremely important.

Propagation in a Hospital-based Wireless PCS

The effective design of a radio network requires accurate description of the channel. One of the most important and difficult issues in wireless communications has been modeling the radio channel. In routine healthcare applications, communication is essentially indoors and involves short distances; therefore, factors such as detailed layout structure, building type, and materials used have to be considered. These factors significantly affect radio propagation and corrections must be made in the existing propagation models for indoor radio communications.

These models are also used to determine the optimum location for the installation of antennas and to analyze the interference between different systems. Hospital construction can highly distort radio wave propagation because multiple reflections of radio waves occur due to the interiors of the building. The received signal will therefore be affected by multi-path as it equals the sum of the attenuated, time delayed, and phase offset versions of the original transmitted signals. With current higher data rates of transmission, intersymbol interference (ISI) of the signal will take place. Prudent planning is necessary for proper signal coverage.

Engineering Solutions for Propagation Issues

As the distance of the receiver from the transmission facility increases, the received signal strength decreases. In a highly cluttered environment like a hospital, there is no direct line-of-sight (LOS) component of the signal between the transmitter and receiver. Therefore, the signals received are highly attenuated. This attenuation causes large scale path loss, and, due to the various reflected components of the transmitted signal, short-term fading over small distances also occurs. Extensive models for the geographical layout of the hospital, contours of equal path loss in the structure of the hospital, shadowing, and diffraction can be used to determine the construction of the medical facility.

Wireless Security in Healthcare

Assuring the privacy of an individual's healthcare data is a key issue in healthcare systems. Even though patient record data can lead to important information for healthcare providers and their patients, there is also a potential for personal harm if disclosed inappropriately. As the data is transferred across wide areas, the integrity of the data must be maintained and the originators and requesters of the data must be confirmed.

Unique identification of patient information is possible through retinal eye scan images, fingerprint readers, DNA blood typing, or personal identification numbers in digitized form. Security of information is important in any communication system and, in healthcare, patient related data needs to be secure and must be available to the concerned medical personnel only. Data encryption and cryptography are methods proposed towards achieving security of information. There are various data encryption standards available that can be used effectively for security in healthcare.

Wireless System Security Design Issues

The role of a wireless system in hospitals is to enable access to patient-related data by the highly mobile physician. Handheld data communicators with the graphical user interface (GUI) software systems will access the centralized computerized patient record (CCPR). The CCPR is, in turn, linked to various hospital departments and other medical facilities. A wireless network accessed through a multimedia handheld terminal (HMT) must have systems reliability and security.

Wireless systems must have system reliability so that accuracy and dependability of data collection, processing, and maintenance are secured through appropriate system design. This includes the use of physical security measures that are directed toward the protection of the environment and equipment. Although currently wired clinical systems are designed to assure 100 percent backup in most environments, the effectiveness of data security in wireless systems has not been fully evaluated. Some of the key security issues in wireless systems are addressed in the following list:

System security: System security is imperative in the wireless healthcare environment. Protection is necessary to prevent unauthorized access.

Data Security: The protection of data from accidental or intentional disclosure to unauthorized persons and from unauthorized alteration. Techniques for security include software and hardware features and data encryption with frequency hop.

Data Integrity: The soundness or completeness of the data that is being used. Data integrity may be maintained by implementing security measures and procedural control, by assigning responsibility, or by establishing audit trails.

Usage Integrity: Implement protection measures against unauthorized access to programs and data, including measures against unintentional or deliberate misuse of patient related data.

Summary

Enabling wireless access to patient related data in the fast-paced healthcare setting by handheld multimedia terminals can enhance the performance and flexibility of the user or physician.

Furthermore, the quality of healthcare delivered improves greatly because data is readily available for diagnosis and interpretation. In addition, cost containment may justify the deployment of such a system. Engineering issues, such as propagation effects, time management, and security, must be considered for effective performance of wireless networks. Environmental safety is another key factor that must be addressed; health risks associated with radio frequency (RF) radiation and pulsed microwave radiation must be considered.

Glossary

Symbols

10Base-2 IEEE standard (known as *thin ethernet*) for 10 Mbps baseband ethernet over coaxial cable at a maximum distance of 185 meters.

10Base-5 IEEE standard (known as *thick ethernet*) for 10 Mbps baseband ethernet over coaxial cable at a maximum distance of 500 meters.

10Base-F IEEE standard for 10 Mbps baseband ethernet over optical fiber.

10Broad-36 IEEE standard for 10 Mbps broadband ethernet over broadband cable at a maximum distance of 3600 meters.

100Base-T IEEE standard for a 100 Mbps LAN.

802.11 IEEE standard for wireless LANs that specifies Medium Access Control (MAC) and Physical Layer (PHY) specifications for 1 Mbps wireless connectivity between fixed, portable, and moving stations within a local area.

A

Acceptance Testing Type of testing that determines whether the network is acceptable to the actual users. The users of the network should participate in developing acceptance criteria and running the tests.

Access Point An interface between the wireless network and a wired network such as ethernet. Access points combined with a wired backbone support the creation of radio cells that facilitate roaming.

Adaptive Routing A form of network routing whereby the path data packets traverse from a source to a destination node depending on the current state of the network. Normally with adaptive routing, routing information stored at each node changes according to some algorithm that calculates the best paths through the network.

ADCCP (Advanced Data Communications Control Procedures) ANSI standard bit-oriented communications protocol.

Agents Software modules that reside in network elements. Agents collect and store management information such as the number of error packets received by a network element.

Algorithm A finite process for the solution to a problem.

ALOHANET Computer sites allowed at seven campuses spread over four islands to communicate with the central computer on Oahu without using existing unreliable and expensive telephone lines.

Alpha Test Product testing done by the vendor.

Analog Cellular A telephone system that uses radio cells to provide connectivity among cellular phones. The analog cellular telephone system uses FM (Frequency Modulation) radio waves to transmit voice grade signals. To accommodate mobility, this cellular system will switch your radio connection from one cell to another as you move between areas. Every cell within the network has a transmission tower that links mobile callers to a Mobile Telephone Switching Office (MTSO).

Analog Signal An electrical signal with an amplitude that varies continuously as time progresses.

Analyst A person capable of gathering information and defining the needs of the users and the organization.

ANSI (American National Standards Institute) The primary standards-forming body in the United States.

ANSI FDDI (Fiber Distributed Data Interface) An ANSI standard for token passing networks. FDDI uses optical fiber and operates at 100 Mbps.

ANSI X.12 An ANSI standard for EDI.

APD (Avalanche Photodiode) A high performance device used in an optical communication system that converts light into electrical signals. Normally, the APD is used in conjunction with an injection laser diode light source.

Appliance Runs applications and is a visual interface between the user and the network. There are several classes of user appliances—the desktop workstation, laptop, palmtop, pen-based computer, Personal Digital Assistant (PDA), and pager.

Application Layer Establishes communications with other users and provides services such as file transfer and electronic mail to the end users of the network.

Application Process An entity, either human or software, that uses the services offered by the Application Layer of the OSI Reference Model.

Application Requirements Specify what specific applications users require.

Application Software Accomplishes the tasks users require, such as word processing, database access, and electronic mail. Therefore, application software directly satisfies network requirements, particularly user requirements.

ARDIS A company that leases access to its wireless WAN that is based on packet radio technology.

ARP (Address Resolution Protocol) A TCP/IP protocol that binds logical (IP) addresses to physical addresses.

ARPANET (Advanced Research Project Agency Network) A Department of Defense network that provided the groundwork for development of the Internet.

ARQ (Automatic Repeat-Request) A method of error correction where the receiving node detects errors and uses a feedback path to the sender for requesting the retransmission of incorrect frames.

Asynchronous Transmission Type of synchronization where there is no defined time relationship between transmission of frames.

AT Bus A PC bus that supports 16 bit-data.

ATM (Asynchronous Transfer Mode) A cell-based connection-oriented data service offering high speed (up to Gbps range) data transfer. ATM integrates circuit and packet switching to handle both constant and burst information.

AUI (Attachment Unit Interface) A 15-pin interface between an ethernet network interface card and transceiver.

B

Bandwidth Specifies the amount of the frequency spectrum that is usable for data transfer. In other words, it identifies the maximum rate a signal can fluctuate without encountering significant attenuation (loss of power).

Baseband Signal A signal that has not undergone any shift in frequency. Normally with LANs, a baseband signal is purely digital.

Baud Rate The unit of signaling speed derived from the duration of the shortest code element of the digital signal. Baud rate is the speed the digital signal pulses travel.

Beta Testing Product testing done by potential users.

Bit Rate The transmission rate of binary digits. Bit rate is equal to the total number of bits transmitted in relation to the time it takes to send them.

Bridge A network component that provides internetworking functionality at the data link or medium access layer of a network's architecture. Bridges can provide segmentation of data frames.

Bus Topology A type of topology where all nodes share a common transmission medium.

C

Cabling Diagram Part of a design specification, illustrating the location of network hardware and the layout of cabling throughout the facility.

CAD (Computer Aided Design) Software Applications that use vector graphics to create complex drawings.

Carrier Current LAN A LAN that uses power lines within the facility as a medium for the transport of data.

Category 1 Twisted-Pair Wire Old-style phone wire, which is not suitable for most data transmission. This includes most telephone wire installed before 1983, in addition to most current residential telephone wiring.

Category 2 Twisted-Pair Wire Certified for data rates up to 4 Mbps, which facilitates IEEE 802.5 Token-Ring networks (4 Mbps version).

Category 3 Twisted-Pair Wire Certified for data rates up to 10 Mbps, which facilitates IEEE 802.3 10Base-T (ethernet) networks.

Category 4 Twisted-Pair Wire Certified for data rates up to 16 Mbps, which facilitates IEEE 802.5 Token-Ring networks (16 Mbps version).

Category 5 Twisted-Pair Wire Certified for data rates up to 100 Mbps, which facilitates ANSI FDDI Token-Ring networks.

CCITT (International Telegraph and Telephone Consultative Committee) An international standards organization that is part of the ITU and dedicated to establishing effective and compatible telecommunications among members of the United Nations. CCITT develops the widely used V-series and X-series standards and protocols.

CDDI (Copper Data Distributed Interface) A version of FDDI specifying the use of unshielded twisted-pair wiring (Category 5).

CDLC (Cellular Data Link Control) Public domain data communications protocol used in cellular telephone systems.

CDPD (Cellular Digital Packet Data) Overlays the conventional analog cellular telephone system, using a channel-hopping technique to transmit data in short bursts during idle times in cellular channels. CDPD operates full duplex in the 800 and 900 Mhz frequency bands, offering data rates up to 19.2 Kbps.

CDRH (Center for Devices and Radiological Health) The part of the U.S. Food and Drug Administration that evaluates and certifies laser products for public use.

Centronics A de facto standard 36-pin parallel 200 Kbps asynchronous interface for connecting printers and other devices to a computer.

CGA (Color/Graphics Adapter) An IBM video display standard providing low-resolution text and graphics.

CGM (Computer Graphics Metafile) A standard format for interchanging graphics images.

Channel Link Consists of the user patch cords, horizontal cross connect, and horizontal wiring. A channel link can span a total of 100 meters.

CLNP (Connectionless Network Protocol) An OSI protocol for providing the OSI Connectionless Network Service (datagram service). CLNP is similar to IP.

CMIP (Common Management Information Protocol) An ISO network monitoring and control standard.

CMIS (Common Management Information Services) An OSI standard defining functions for network monitoring and control.

CMOL (CMIP over LLC) A version of CMIP that runs on IEEE 802 LANs.

CMOT (CMIP over TCP) A version of CMIP that runs on TCP/IP networks.

Coaxial Cable Type of medium having a solid metallic core with a shielding as a return path for current flow. The shielding within the coaxial cable reduces the amount of electrical noise interference within the core wire; therefore, coaxial cable can extend to much greater lengths than twisted-pair wiring.

Constraints Requirements that are impossible or infeasible to change. Constraints limit the project team's options in completing the project. Common constraints are amount of funding, technical requirements, availability of resources, type and location of project staff, and schedules.

Communications Cabinet Drawings Illustrate the placement of network components that require mounting within communications cabinets.

Concept of Operations Defines system-level functionality, operational environment, and implementation priorities of an information system. Project team members use the concept of operations as a basis for planning the development and determining the requirements of the information system.

Connectivity A path for communications signals to flow through. Connectivity exists between a pair of nodes if the destination node can correctly receive data from the source node at a specified minimum data rate.

Customer Focal Point Represents the interests of the people who will be using the network. The customer focal point aims the project team in the right direction when gathering information determining requirements.

D

Datagram Service A connectionless form of packet switching whereby the source does not need to establish a connection with the destination before sending data packets.

DB-9 A standard 9-pin connector commonly used with RS-232 serial interfaces on portable computers. The DB-9 connector will not support all RS-232 functions.

DB-15 A standard 15-pin connector commonly used with RS-232 serial interfaces, ethernet transceivers, and computer monitors.

DB-25 A standard 25-pin connector commonly used with RS-232 serial interfaces. The DB-25 connector will support all RS-232 functions.

DES (Data Encryption Standard) A cryptographic algorithm that protects unclassified computer data. DES is a National Institute of Standards and Technology (NIST) standard and is available for both public and government use.

Design Process that determines how the network will meet requirements. Design involves the selection of technologies, standards, and products that provide a solution to the stated requirements.

Design Phase Consists of selecting a set of technologies, standards, and products that satisfy requirements.

Design Specifications Describe the technologies, standards, components, and configurations of hardware and software comprising the network.

Desktop Conferencing An application enabling users to have televideo conferences directly from PCs located in their offices. Users who participate in a desktop conference can hear and see a video image of each participant. Desktop conferences also enable participants to jointly edit documents and facilitate electronic white boards.

Desktop Publishing Software Applications that provide the ability to effectively merge text and graphics and maintain precise control of the layout of each page of the document.

Detailed Design The final phase of network design that selects components and finalizes network configurations and documentation.

Diffused Laser Light Type of laser transmission where the light is reflected off a wall or ceiling.

Direct Sequence Spread Spectrum Combines a data signal at the sending station with a higher data rate bit sequence, which many refer to as a chip sequence (also known as processing gain). A high processing gain increases the signal's resistance to interference. The minimum processing gain that the FCC allows is 10, and most products operate under 20.

Directional Antenna Type of antenna that sends radio waves primarily in one direction.

Distributed Routing A form of routing where each node (router) in the network periodically identifies neighboring nodes, updates its routing table, and, with this information, then sends its routing table to all of its neighbors. Because each node follows the same process, complete network topology information propagates through the network and eventually reaches each node.

Document Management An application enabling users to effectively manage and access information contained within different file types. Most document management software enables users to manipulate both data and images, as well as easily manage files through indexing, search, and query functions.

DQDB (Distributed Queue Dual Bus) A technology that provides full duplex 155 Mbps operation between nodes of a metropolitan area network. The IEEE 802.6 standard is based on DQDB.

DSU/CSU (Data Service Unit/Channel Service Unit) A network component that reshapes data signals into a form that can be effectively transmitted over a digital transmission medium, typically a leased 56 Kbps or T1 line.

Dynamic Host Configuration Protocol (DHCP) Issues IP addresses automatically within a specified range to devices such as PCs when they are first powered on. The device retains the use of the IP address for a specific license period that the system administrator can define. DHCP is available as part of the Microsoft Windows NT Server network operating system.

E

EGA (Enhanced Graphics Adapter) An IBM video display standard providing medium-resolution text and graphics.

EIA (Electronics Industry Association) A domestic standards-forming organization that represents a vast number of United States electronics firms.

EISA (Extended ISA) A PC bus standard that extends the 16-bit ISA bus to 32 bits and runs at 8 MHz.

Electronic Data Interchange (EDI) A service that provides standardized inter-company computer communications for business transactions. ANSI standard X.12 defines the data format for business transactions for EDI.

Environmental requirements Report conditions, such as weather, pollution, presence and intensity of electromagnetic waves, building construction, and floor space that could affect the operation of the system.

Ethernet A 10 Mbps LAN medium-access method that uses CSMA to allow the sharing of a bus-type network. IEEE 802.3 is a standard that specifies ethernet.

F

FDDI (Fiber Distributed Data Interface) An ANSI standard for token-passing networks. FDDI uses optical fiber and operates at 100 Mbps.

FEC (Forward Error Correction) A method of error control where the receiving node automatically corrects as many channel errors as it can without referring to the sending node.

File Transfer Capability to send files from one computer to another. File transfer mechanisms enable you to move a file from a disk directory on one computer to that of another. The File Transfer Protocol (FTP), a de facto standard, is a popular protocol for transferring files. Issues involved with file transfer include file size and type. The larger the file, the more bandwidth it takes to deliver, possibly making it infeasible to send very large files. Some networks will only support the transfer of text-only files, but other networks are capable of sending binary files as well.

Firewall A device that interfaces the network to the outside world and shields the network from unauthorized users. The firewall does this by blocking certain types of traffic. For example, some firewalls permit only electronic mail traffic to enter the network from elsewhere. This helps protect the network against attacks made to other network resources, such as sensitive files, databases, and applications.

Fractional T-1 A 64 Kbps increment of a T1 frame.

Frame Relay A packet-switching interface that operates at data rates of 56 Kbps to 2 Mbps. Actually, frame relay is similar to X.25, minus the transmission error control overhead. Thus, frame relay assumes that a higher layer, end-to-end protocol will check for transmission errors. Carriers offer frame relay as permanent connection-oriented (virtual circuit) service.

Frequency-Hopping Spread Spectrum Takes the data signal and modulates it with a carrier signal that hops from frequency to frequency as a function of time over a wide band of frequencies. For example, a frequency-hopping radio will hop the carrier frequency over the 2.4 GHz frequency band between 2.4 GHz and 2.483 GHz. A hopping code determines the frequencies it will transmit and in which order. To properly receive the signal, the receiver must be set to the same hopping code and "listen" to the incoming signal at the right time at the correct frequency.

FTAM (File Transfer, Access, and Management) An OSI remote file service protocol.

FTP (File Transfer Protocol) A TCP/IP protocol for file transfer.

Fully-Connected Topology A topology where every node is directly connected to every other node in the network.

Functional Requirements Describe what the users and organization expect the system to do. Therefore, functional requirements closely align to the tasks and actions users perform.

G

Gateway A network component that provides interconnectivity at higher network layers. For example, electronic mail gateways can interconnect dissimilar electronic mail systems.

Global Positioning System (GPS) a worldwide, satellite-based radio navigation system providing three-dimensional position, velocity and time information to users having GPS receivers anywhere on or near the surface of the Earth.

H

Hardware Physical elements containing the electronics necessary to move, store, and display information on the network. The hardware provides connectivity between the users and information stored in memory devices. Hardware, such as monitors and printers, provides a means to view the information. Other examples of network hardware elements include network interface cards, server and client platforms, cabling, bridges, and routers.

Hardware Configuration Plan Identifies how the team should configure hardware the network utilizes.

Hardware Platform A computer that provides necessary processors and file storage. The processors execute the program code, enabling the software to run. The file storage facilitates the storage of software programs and data files. With a network, applications and network operating systems normally reside on some type of hardware platform.

HDLC (High-level Data Link Control) An ISO bit-oriented protocol for link synchronization and error control.

Help Desk A central point of contact for users needing assistance when using the network and its resources.

Hierarchical Topology A topology where nodes in the same geographical area are joined together, then tied to the remaining network as groups. The idea of a hierarchical topology is to install more links within high density areas and fewer links between these populations.

HSSI (High-Speed Serial Interface) A standard for up to 52 Mbps serial connections, often used to connect T3 lines.

HTML (HyperText Markup Language) A standard used on the Internet World Wide Web for defining hypertext links between documents.

I

IEEE (Institute of Electrical and Electronic Engineers) A United States-based standards organization participating in the development of standards for data transmission systems. IEEE has made significant progress in the establishment of standards for LANs, namely the IEEE 802 series of standards.

IEEE 802.1 IEEE standard for network management.

IEEE 802.1D IEEE standard for inter-LAN bridges.

IEEE 802.2 IEEE standard for Logical Link Control (LLC).

IEEE 802.3 IEEE standard for CSMA/CD (ethernet) LAN access.

IEEE 802.4 IEEE standard for token-bus LAN multiple access.

IEEE 802.5 IEEE standard for token-ring LAN multiple access.

IEEE 802.6 IEEE standard for DQDB metropolitan area network multiple access.

IEEE 802.7 IEEE standard for broadband LANs.

IEEE 802.9 IEEE standard for integrated digital and video networking.

IEEE 802.10 IEEE standard for network security.

IEEE 802.11 IEEE standard for wireless LANs.

IEEE 488 IEEE standard for computer-to-electronic instrument communication.

IEEE 802.12 IEEE standard for a demand priority LAN access (also called *Fast Ethernet*).

IEEE 1284 IEEE standard for an enhanced parallel port compatible with the Centronics parallel port.

Information Flow Requirements Specify the paths of information flow between people and system, types and formats of information sent, frequency of information transmission, and maximum allowable error rates.

Infoware Applications that provide online encyclopedias, magazines, and other references.

Infrared Light Light waves having wavelengths ranging from about 0.75 to 1,000 microns, which is longer (lower in frequency) than the spectral colors but much shorter (higher in frequency) than radio waves. Therefore, under most lighting conditions, infrared light is invisible to the naked eye.

Infrared Light LAN A LAN that uses infrared light as its medium.

Installation Plan Describes the tools and procedures necessary for an installation team to install and test network hardware and software components.

Instructor Guide Explains how to teach a particular course. This is especially useful if the instructors are people who were not involved in the course development.

Integration Testing Type of testing that verifies the interfaces between network components as the components are installed. The installation crew should integrate components into the network one-by-one and perform integration testing when necessary to ensure proper gradual integration of components.

Internetwork A collection of interconnected networks. Often it is necessary to connect networks together, and an internetwork provides the connection between different networks. One organization having a network may want to share information with another organization having a different network. The internetwork provides functionality needed to share information between these two networks.

Internetworking A mechanism defining the communications process necessary to connect two dissimilar autonomous networks.

Inward Interference Interference coming from other devices, such as microwave ovens and other wireless network devices, that will result in delay to the user by either blocking transmissions from stations on the LAN, or by causing bit errors to occur in data being sent.

ISA (Industry Standard Architecture) A widely used PC expansion bus that accepts plug-in boards, such as ethernet cards, video display boards, and disk controllers. ISA was originally called the AT bus because it was first used in the IBM AT, extending the original bus from eight to 16 bits. Most ISA PCs provides a mix of 8-bit and 16-bit expansion slots. Contrast with EISA and Micro Channel.

ISDN (Integrated Services Digital Network) A collection of CCITT standards specifying WAN digital transmission service. The overall goal of ISDN is to provide a single physical network outlet and transport mechanism for the transmission of all types of information, including data, video, and voice.

IS-IS (Intermediate System to Intermediate System) Protocol An OSI protocol for intermediate systems exchange routing information.

ISM (Industrial, Scientific, and Medicine) Radio frequency bands that the Federal Communications Commission (FCC) authorized for wireless LANs. The ISM bands are located at 902 MHz, 2.400 GHz, and 5.7 GHz.

ISO (International Standards Organization) A non-treaty standards organization active in the development of international standards such as the Open System Interconnection (OSI) network architecture.

ISO 7498 ISO standard for the Open System Interconnection (OSI) basic reference model.

ISO 9001 ISO standards for quality design, development, production, installation, and service procedures.

Isochronous Transmission Type of synchronization where information frames are sent at specific times.

ITU (International Telecommunications Union) An agency of the United States providing coordination for the development of international standards.

J

Joint Application Design (JAD) A parallel process simultaneously defining requirements in the eyes of the customer, users, sales people, marketing staff, project managers, analysts, and engineers. You can use the members of this team to define requirements.

JPEG (Joint Photographic Experts Group) An ISO standard for lossy compression.

L

Lantastic A peer-to-peer network operating system by Artisoft, Inc.

LAP (Link Access Procedure) An ITU error correction protocol derived from the HDLC standard.

LAP-B (LAP-Balanced) LAP protocol used in X.25 networks.

LAP-D (LAP-D channel) LAP protocol used in the data channel of ISDN networks.

LAP-X (LAP-Half-dupleX) LAP protocol used for ship-to-shore transmission.

LASER Is a common term for Light Amplification by Stimulated Emission of Radiation, a device containing a substance where the majority of its atoms or molecules are put into an excited energy state. As a result, the laser emits coherent light of a precise wavelength in a narrow beam. Most laser MANs use lasers that produce infrared light.

Learning Objectives A training development element specifying what students are expected to do after receiving instruction.

LED (Light Emitting Diode) Used in conjunction with optical fiber, it emits incoherent light when current is passed through it. Advantages to LEDs include low cost and long lifetime, and they are capable of operating in the Mbps range.

Local Area Network (LAN) A computer network confined to a local area, such as a building. The specific uses of a LAN include inter-office communications and peripheral sharing. Usually, a LAN is owned by a single funding organization. That is, a single department, division or organization will normally independently purchase and install the LAN hardware and software.

Local Bridge A bridge that connects two LANs within close proximity.

M

Mail Gateway A type of gateway that interconnects dissimilar electronic mail systems.

Maintenance An operational support function that performs preventative maintenance on the network and troubleshoots and repairs the network if it becomes inoperable.

Management Information Base (MIB) A collection of managed objects residing in a virtual information store.

MAP (Manufacturing Automation Protocol) A protocol used extensively by General Motors for automated factory floor equipment.

MAU (Multi-station Access Unit) A multiport wiring hub for token-ring networks.

MCA (Micro Channel Architecture) An internal 32-bit bus originally introduced by IBM.

Medium A physical link that provides a basic building block to support the transmission of information signals. Most media are composed of either metal, glass, plastic, or air.

Medium Access A data link function that controls the use of a common network medium.

Meteor Burst Communications A communications system that directs a radio wave, modulated with a data signal, at the ionosphere. The radio signal reflects off the ionized gas left by the burning of meteors entering the atmosphere and is directed back to Earth in the form of a large footprint, enabling long distance operation.

MHS (Message Handling Service) A Novell messaging system that supports multiple operating systems and messaging protocols.

MIB (Management Information Base) A collection of objects that can be accessed via a network management protocol.

Microsoft Windows NT Server A server-oriented network operating system offering excellent support of applications, as well as file and print services.

MIPS (Million Instructions Per Second) Identifies the number of instructions a computer processes over time. High-speed personal computers, such as pentiums, are usually capable of operating at 100 MIPS or greater. A 386 PC usually runs between 3 to 5 MIPS. However, MIPS rates are not uniform because some vendors use the best case value of the platform and others use average rates.

MIDI (Musical Instrument Digital Interface) A standard protocol for the interchange of musical information between musical instruments and computers.

Mobility Ability to continually move from one location to another.

Mobility Requirements Describe the movement of the users when performing their tasks. Mobility requirements should distinguish whether the degree of movement is continuous or periodic.

Modulation The process of translating the baseband digital signal to a suitable analog form.

Motif A standard GUI for Unix. Motif is endorsed by the Open Software Foundation.

MPEG (Moving Pictures Experts Group) An ISO standard for lossless compression of full-motion video.

Multimedia The integration of graphics, text, and sound into a single application.

Multimedia Software Applications that add graphics, sound, and video for use in education and specialized applications.

Multiplexer A network component that combines multiple signals into one composite signal in a form suitable for transmission over a long-haul connection, such as leased 56 Kbps or T1 circuits.

N

Narrowband System A wireless system that uses dedicated frequencies assigned by the FCC licenses. The advantage of narrowband systems is that if interference occurs, the FCC will intervene and issue an order for the interfering source to cease operations. This is especially important when operating wireless MANs in areas having a great deal of other operating radio-based systems.

NetBEUI A protocol that governs data exchange and network access. NetBEUI is Microsoft's version of NetBIOS.

NetBIOS (Network Basic Input Output System) A standard interface between networks and PCs.

NetWare A server-based network operating system by Novell, Inc.

NetWare Loadable Module (NLM) An application co-existing with the core NetWare network operating system.

Network Documentation A description of the network that identifies and describes all hardware, software, protocols, people, and operating environments. Complete documentation identifies a baseline that describes the current makeup of the system. Documentation is very important because it allows people to effectively use and maintain the system.

Network Interface Card (NIC) An external modem that facilitates the modulator and communications protocols.

Network Layer Provides the routing of packets from source to destination.

Network Management Consists of a variety of elements that protect the network from disruption and provide proactive control of the configuration of the network.

Network Management Station (NMS) Executes management applications that monitor and control network elements.

Network Monitoring A form of operational support enabling network management to view the inner-workings of the network. Most network monitoring equipment is non-obtrusive and can determine the network's utilization and locate faults.

Network Operating System (NOS) A network component that provides a platform for network applications to operate. A NOS normally offers communications, printing, and file services for applications residing on the network.

Network Re-engineering A structured process that can help an organization proactively control the evolution of its network. Network re-engineering consists of continually identifying factors influencing network changes, analyzing network modification feasibility, and performing network modifications as necessary.

Network Requirements Define detailed functional and system requirements, operational environment, and constraints of the network.

Network Security A form of operational support that attempts to protect the network from compromise and destruction, as well as make sure the network will be available when needed. Network security includes elements such as access control, data encryption, and data backup.

NFS (Network File System) A distributed file system enabling a set of dissimilar computers to access each other's files in a transparent manner.

Node Any network-addressable device on the network, such as a router or network interface card.

NSAP (Network Service Access Point) A point in the network where OSI network services are available to a transport entity.

0

ODI (Open Data-Link Interface) Novell's specification for network interface card device drivers, allowing simultaneous operation of multiple protocol stacks.

Office Software Applications that typical office personnel use, such as word processing, databases, spreadsheets, graphics, and electronic mail.

Omnidirectional Antenna Type of antenna that sends radio waves in all directions.

Operating Environment The physical location where the network will operate.

Operational Support Plan Describes how the organization will support the operational network in terms of system administration, network monitoring and control, accounting and chargeback, maintenance, security, configuration management, training, and system re-engineering.

Operational Support Preparation Phase The planning necessary to effectively support the system after it is installed. Preparations include training development and delivery, and plans for support elements such as maintenance, system administration, and security.

Operational Support Requirements Define any elements needed to effectively integrate the system into the existing operational support infrastructure. For example, you should require the inclusion of Simple Network Management Protocol (SNMP) if current network monitoring stations require SNMP.

Operator Profiles Identify and describe attributes of each person who will be operating the system.

Optical Fiber A type of medium that uses changes in light intensity to carry information through glass or plastic fibers.

OSI (Open System Interconnection) An ISO standard specifying an open system capable of enabling the communications between diverse systems. OSI has seven layers of distinction. These layers provide the functions necessary to allow standardized communications between two application processes.

OSPF (Open Shortest Path First) Routing protocol for TCP/IP routers.

P

Packet Radio Uses packet switching to move data from one location to another across radio links.

PCM (Pulse Code Modulation) A common method for converting analog voice signals into a digital bit stream.

PCMCIA (Personal Computer Memory Card International Association) A standard set of physical interfaces for portable computers. PCMCIA specifies three interface sizes—Type I (3.3 millimeters), Type II (5.0 millimeters), and Type III (10.5 millimeters).

Peer-to-Peer Network A network where there are communications between a group of equal devices. A peer-to-peer LAN does not depend upon a dedicated server, but allows any node to be installed as a non-dedicated server and share its files and peripherals across the network. Peer-to-peer LANs are normally less expensive because they do not require a dedicated computer to store applications and data. They do not perform well, however, for larger networks.

Performance Modeling The use of simulation software to predict network behavior, allowing you to perform capacity planning. Simulation allows you to model the network and impose varying levels of utilization to observe the effects.

Performance Monitoring Addresses performance of a network during normal operations. Performance monitoring includes real-time monitoring, where metrics are collected and compared against thresholds that can set off alarms; recent-past monitoring, where metrics are collected and analyzed for trends that may lead to performance problems; and historical data analysis, where metrics are collected and stored for later analysis.

Performance Requirements Identify expected values for reliability, availability, and delay.

Peripherals Items such as printers, fax machines, and modems.

Personal Communications Services (PCS) A spectrum allocation located at 1.9 GHz, a new wireless communications technology offering wireless access to the World Wide Web, wireless e-mail, wireless voice mail, and cellular telephone service.

Physical Layer Provides the transmission of bits through a communication channel by defining electrical, mechanical, and procedural specifications.

Polymorphic Virus A virus that, to avoid identification, changes its binary pattern each time it infects a new file.

Portability Defines network connectivity that can be easily established, used, then dismantled.

POSIX (Portable Operating System Interface for Unix) An IEEE 1003.1 standard defining the interface between application programs and the Unix operating system.

POTS (Plain Old Telephone System) The common analog telephone system developed many years ago for voice communications.

PPP (Point-to-Point Protocol) A protocol that provides router-to-router and host-to-network connections over both synchronous and asynchronous circuits. PPP is the successor to SLIP.

Presentation Layer Negotiates data transfer syntax for the application layer and performs translations between different data types, if necessary.

Project Charter Formally recognizes the existence of the project, identifies the business need that the project is addressing, and gives a general description of the resulting product.

Project Management Overseers needed to make sure actions are planned and executed in a structured manner.

Protocols Rules that the system must follow to operate correctly. Most network protocols are based on technology and describe rules for communications among system hardware and software elements. These rules govern format, timing, sequencing, and error control. Without protocols, system elements would not be able to make sense of communications from other elements. Most systems have sets of protocols often referred to as protocol stacks. Many protocols have been established as standards, approved by various national or international organizations. Examples of protocols include ethernet, token-ring, and X.400.

Prototyping A method of determining or verifying requirements and design specifications. The prototype normally consists of network hardware and software that support a proposed solution. The approach to prototyping is typically a trial-and-error experimental process.

R

RAM Mobile Data A company that leases access to its wireless WAN that is based on packet radio technology.

Red Book A document of the United States National Security Agency (NSA) defining criteria for secure networks.

Relay Node Implements a routing protocol that maintains the optimum routes for the routing tables, forwarding packets closer to the destination.

Remote Bridge A bridge that connects networks separated by longer distances. Organizations use leased 56 Kbps circuits, T1 digital circuits, and radio waves to provide long distance connections between remote bridges.

Repeater A network component that provides internetworking functionality at the physical layer of a network's architecture. A repeater regenerates digital signals.

Requirements Identify *what* the network is supposed to do, not *how* it's supposed to do it. Requirements are crucial in all development projects—they provide the basis for design, implementation, and support of the system or product.

Requirements Analysis A process of defining what the network is supposed to do, providing a basis for the network design.

Requirements Phase Defines the needs of the users and the organization of the eventual network or system, providing the basis for a solution.

Resource Requirements Identify people and equipment the project will need to accomplish the project's goals.

Resource Sharing A network service allowing users of the network to share the use of network resources, such as printers, disk drives, applications, modems, and fax machines.

Ring Topology A topology where a set of nodes are joined in a closed loop.

Router A network component that provides internetworking at the network layer of a network's architecture by allowing individual networks to become part of a WAN.

Routing Information Protocol (RIP) A common type of routing protocol. RIP bases its routing path on the distance (number of hops) to the destination. RIP maintains optimum routing paths by sending out routing update messages if the network topology changes. For example, if a router finds that a particular link is faulty, it will update its routing table, then send a copy of the modified table to each of its neighbors.

RS Recommended Standard.

RS-232 An EIA standard for up to 20 Kbps, 50 foot, serial transmission between computers and peripheral devices.

RS-422 An EIA standard specifying electrical characteristics for balanced circuits. RS-422 is used in conjunction with RS-449.

RS-423 An EIA standard specifying electrical characteristics for unbalanced circuits. RS-423 is used in conjunction with RS-449.

RS-449 An EIA standard specifying a 37-pin connector for high-speed transmission.

RS-485 An EIA standard for multipoint communications lines.

S

SAP (Service Access Point) A point at which the services of an OSI layer are made available to the next higher layer.

Schematic A drawing that is normally part of a design specification and illustrates the electrical connections between network hardware components. The schematic serves two main purposes: the installation team uses the schematic to properly install the network, and maintenance technicians use the schematic to facilitate effective troubleshooting.

SCI (Scalable Coherent Interface) An IEEE standard for a high-speed (up to 1 Gbps) bus.

Scientific Applications Applications that provide the analysis of real-world events by simulating them with mathematics.

Security Requirements Identify what information and systems require protection from particular threats. The degree of security requirements depends on the severity of the consequences the organization would face if it lost information or part of the system were destroyed.

Server-Oriented Network A network architecture where the network software is split into two pieces, one each for the client and the server. The server component provides services for the client software; the client part interacts with the user. The client and server components run on different computers, and the server is usually more powerful than the client. The main advantages of a server-oriented network is less network traffic. Therefore, networks having a large number of users will normally perform better with server-oriented networks.

Session Layer Establishes, manages, and terminates sessions between applications.

SIMM (Single In-line Memory Module) Standard packaging for PC memory.

SLIP (Serial Line Internet Protocol) An Internet protocol used to run IP over serial lines.

SMDS (Switched Multimegabit Digital Service) A packet switching connectionless data service for WANs.

SMTP (Simple Mail Transfer Protocol) The Internet electronic mail protocol.

SNA (Systems Network Architecture) IBM's proprietary network architecture.

SNMP (Simple Network Monitoring Protocol) A network management protocol that defines the transfer of information between Management Information Bases (MIBs). Most high-end network monitoring stations require the implementation of SNMP on each of the components the organization wishes to monitor.

Software Configuration Plan Identifies how the team should configure software the network uses.

SONET (Synchronous Optical NETwork) A fiber optic transmission system for high-speed digital traffic. SONET is part of the B-ISDN standard.

Spectrum Analyzer An instrument that identifies the amplitude of signals at various frequencies.

Spread Spectrum A modulation technique that spreads a signal's power over a wide band of frequencies. The main reasons for this technique is that the signal becomes much less susceptible to electrical noise and interferes less with other radio-based systems.

SQL (Structured Query Language) An international standard for defining and accessing relational databases.

ST Connector An optical fiber connector that uses a bayonet plug and socket.

Star Topology A topology where each node is connected to a common central switch or hub.

Statement Of Work (SOW) Describes what needs to be done to accomplish the network modification.

Stealth Virus A virus that is undetectable.

Switched Multimegabit Data Service (SMDS) A packet switching interface that operates at data rates ranging from 1.5 Mbps to 45 Mbps. SMDS is similar to frame relay, except SMDS provides connectionless (datagram) service.

Synchronous Transmission Type of synchronization where information frames are sent within certain time periods.

System Administration Type of operational support that provides a human interface between the system and its users. A system administrator assigns user addresses and log-in passwords, allocates network resources, and performs some network security functions.

System Interface Requirements Describe the architectures of these systems and know the hardware, software, and protocols necessary for proper interfacing.

System Testing Type of testing that verifies the installation of the entire network. Testers normally complete system testing in a simulated production environment, simulating actual users in order to ensure the network meets all stated requirements.

T

T1 A standard specifying a time division multiplexing scheme for point-to-point transmission of digital signals at 1.544 Mbps.

TCP (Transmission Control Protocol) A common standard transport layer protocol.

TDR (Time-Domain Reflectometer) Tests the effectiveness of network cabling.

Technical Interchange Meeting (TIM) Addresses technical issues needing attention by project team members and customer representatives. A TIM is effective if the solution to a technical requirement or problem cannot be adequately solved by a single team member.

Technical Service Bulletin (TSB) 67 Describes how to test Category 5 twisted-pair cable. TSB 67 was published by the Link Performance Task Group, a subcommittee of the Telecommunications Industry Association's TR41 Standards Committee.

Technology Comparison Matrix A documentation method that compares similar technologies based on attributes such as functionality, performance, cost, and maturity.

Telecommuting The concept of electronically stretching an office to a person's home.

Telnet A virtual terminal protocol used in the Internet, enabling users to log into a remote host.

Terminal Node Controllers (TNCs) Interface their computers through ham radio equipment. These TNCs act much like a telephone modem, converting the computer's digital signal into one that a ham radio can modulate and send over the airwaves using a packet switching technique.

Test Case An executable test with a specific set of input values and a corresponding expected result.

Token Ring A medium access method that provides multiple access to a ring type network through the use of a token. FDDI and IEEE 802.5 are token-ring standards.

TOP (Technical and Office Protocol) A protocol used extensively by General Motors to exchange information between the factory floor and engineering offices.

Top-Down Design First defines high-level specifications directly satisfying network requirements, then defines the remaining elements in an order that satisfies the most specifications already determined.

Topography A description of the network's physical surface spots. Topography specifies the type and location of nodes with respect to one another.

Topology A description of the network's geographical layout of nodes and links.

TP0 OSI Transport Protocol Class 0 (Simple Class), useful only with very reliable networks.

TP4 OSI Transport Protocol Class 4 (Error Detection and Recovery Class), useful with any type of network. The functionality of TP4 is similar to TCP.

Transceiver A device for transmitting and receiving packets between the computer and the medium.

Transmission Control Protocol (TCP) A commonly used protocol for establishing and maintaining communications between applications on different computers. TCP provides full-duplex, acknowledged, and flow-controlled service to upper-layer protocols and applications.

Transport Layer Provides mechanisms for the establishment, maintenance, and orderly termination of virtual circuits, while shielding the higher layers from the network implementation details.

Twisted Pair Type of medium using metallic type conductors twisted together to provide a path for current flow. The wire in this medium is twisted in pairs to minimize the electromagnetic interference between one pair and another.

U

Unit Testing Type of testing that verifies the accuracy of each network component, such as servers, cables, hubs, and bridges. The goal of unit testing is to make certain the component works properly by running tests that fully exercise the internal workings of the component.

Unix A multiuser and multitasking operating system originally developed by AT&T.

UPS (Uninterruptible Power Supply) A network component that provides a short supply of power to servers and other critical devices if power discontinues.

User Interface Mechanism that enables users to access network services.

User Manual A document providing user-level procedures on how to log in and utilize network services and applications.

User Profile Requirements Identify the attributes of each person who will be using the system, providing human factors that designers can use to select or develop applications.

V

V.21 An ITU standard for asynchronous 0-300 bps full-duplex modems.

V.21 FAX An ITU standard for facsimile operations at 300 bps.

V.22 An ITU standard for asynchronous and synchronous 600 and 1200 bps full-duplex modems.

V.22 bis An ITU standard for 2,400 bps duplex modems.

V.23 An ITU standard for asynchronous and synchronous 0-600 and 0-1200 bps half-duplex modems.

V.32 An ITU standard for asynchronous and synchronous 4800 and 9600 bps full-duplex modems.

V.32terbo An AT&T standard for 19,200 bps modems.

V.33 An ITU standard for synchronous 12,000 and 14,400 bps full-duplex modems.

V.34 An ITU standard for 28,800 bps modems.

V.35 An ITU standard for group band modems combining several telephone circuits to achieve high data rates.

V.42 An ITU standard for modem error checking.

Vertical Market Applications Applications that provide customized data entry, query, and report functions for various industries, such as insurance and banking.

VGA (Video Graphics Array) An IBM video display standard.

VINES A server-oriented network operating system by Banyan, Inc.

Virus Software that infects a computer and causes it to malfunction and destroy data.

W

Wide Area Network (WAN) A type of network that makes long-haul, inter-office communications possible. WANs provide information transport between LANs and users, often over a wide geographical area.

Wireless MAN Provides communications links between buildings, avoiding the costly installation of cabling or leasing fees and the down time associated with system failures.

Wireless Network Interface Couples the digital signal from the end-user appliance to the wireless medium, which is air.

Wiremap Test Ensures a link has proper connectivity by testing for continuity and other installation mistakes, such as the connection of wires to the wrong connector pin.

Work Breakdown Structure (WBS) Shows how the team will accomplish the project by listing all tasks the team will need to perform and the products they must deliver.

World Wide Web An interconnection of privately owned and operated servers. Each server stores a set of hypertext pages that users, or "Web Surfers," can easily search for and view from their workstations through a Web browser via an Internet connection. Also known as *the Web*.

X

X.21 An ITU standard for a circuit switching network.

X.25 An ITU standard for an interface between a terminal and a packet switching network. X.25 was the first public packet switching technology, developed by the CCITT and offered as a service during the 1970s and still available today. X.25 offers connection-oriented (virtual circuit) service and operates at 64 Kbps, which is too slow for some high-speed applications.

X.75 An ITU standard for packet switching between public networks.

X.121 An ITU standard for international address numbering.

X.400 An ITU standard for OSI messaging.

X.500 An ITU standard for OSI directory services.

X Windows A windowing system that runs under Unix.

INDEX

C

J–K–L

T

X–Y–Z

WANT MORE INFORMATION?

CHECK OUT THESE RELATED TOPICS OR SEE YOUR LOCAL BOOKSTORE

CAD and 3D Studio

As the number one CAD publisher in the world, and as a Registered Publisher of Autodesk, New Riders Publishing provides unequaled content on this complex topic. Industry-leading products include AutoCAD and 3D Studio.

Networking

As the leading Novell NetWare publisher, New Riders Publishing delivers cutting-edge products for network professionals. We publish books for all levels of users, from those wanting to gain NetWare Certification, to those administering or installing a network. Leading books in this category include *Inside NetWare 3.12*, *CNE Training Guide: Managing NetWare Systems*, *Inside TCP/IP*, and *NetWare: The Professional Reference*.

Graphics

New Riders provides readers with the most comprehensive product tutorials and references available for the graphics market. Best-sellers include *Inside CorelDRAW! 5*, *Inside Photoshop 3*, and *Adobe Photoshop NOW!*

Internet and Communications

As one of the fastest growing publishers in the communications market, New Riders provides unparalleled information and detail on this ever-changing topic area. We publish international best-sellers such as *New Riders' Official Internet Yellow Pages, 2nd Edition*, a directory of over 10,000 listings of Internet sites and resources from around the world, and *Riding the Internet Highway, Deluxe Edition*.

Operating Systems

Expanding off our expertise in technical markets, and driven by the needs of the computing and business professional, New Riders offers comprehensive references for experienced and advanced users of today's most popular operating systems, including *Understanding Windows 95*, *Inside Unix*, *Inside Windows 3.11 Platinum Edition*, *Inside OS/2 Warp Version 3*, and *Inside MS-DOS 6.22*.

Other Markets

Professionals looking to increase productivity and maximize the potential of their software and hardware should spend time discovering our line of products for Word, Excel, and Lotus 1-2-3. These titles include *Inside Word 6 for Windows*, *Inside Excel 5 for Windows*, *Inside 1-2-3 Release 5*, and *Inside WordPerfect for Windows*.

Orders/Customer Service 1-800-653-6156   Source Code NRP95

New Riders Publishing   201 West 103rd Street ◆ Indianapolis, Indiana 46290 USA

PLUG YOURSELF INTO...

The Macmillan USA Information SuperLibrary (tm)

See the new SuperLibrary Newsletter

THE MACMILLAN
INFORMATION SUPERLIBRARY™

Free information and vast computer resources from the world's leading computer book publisher—online!

FIND THE BOOKS THAT ARE RIGHT FOR YOU!
A complete online catalog, plus sample chapters and tables of contents!

- **STAY INFORMED** with the latest computer industry news through our online newsletter, press releases, and customized Information SuperLibrary Reports.

- **GET FAST ANSWERS** to your questions about Macmillan Computer Publishing books.

- **VISIT** our online bookstore for the latest information and editions!

- **COMMUNICATE** with our expert authors through e-mail and conferences.

- **DOWNLOAD SOFTWARE** from the immense Macmillan Computer Publishing library:
  - Source code, shareware, freeware, and demos

- **DISCOVER HOT SPOTS** on other parts of the Internet.

- **WIN BOOKS** in ongoing contests and giveaways!

TO PLUG INTO MCP:

WORLD WIDE WEB: **http://www.mcp.com**

FTP: ftp.mcp.com

Name _____ Title _____

Company _____ Type of business _____

Address _____

City/State/ZIP _____

Have you used these types of books before?   ☐ yes   ☐ no

If yes, which ones? _____

_____

How many computer books do you purchase each year?   ☐ 1–5   ☐ 6 or more

How did you learn about this book? _____

Where did you purchase this book? _____

Which applications do you currently use? _____

_____

Which computer magazines do you subscribe to? _____

_____

What trade shows do you attend? _____

_____

Comments: _____

_____

_____

Would you like to be placed on our preferred mailing list?   ☐ yes   ☐ no

---

☐ **I would like to see my name in print!** You may use my name and quote me in future New Riders products and promotions. My daytime phone number is: _____

New Riders Publishing   201 West 103rd Street   ◆   Indianapolis, Indiana 46290 USA

Fax to **317-581-4670**   Orders/Customer Service **1-800-653-6156**   Source Code **NRP95**

Fold Here

NEW RIDERS PUBLISHING
201 W 103RD ST
INDIANAPOLIS IN 46290-9058